THE PRINCIPLES AND PRACTICE OF ELECTRICAL EPILATION

This book is dedicated to Peter, a special friend whose inspiration has been invaluable; and Rod Fabes, who sadly died in March 1995. His support and encouragement in the field of epilation was constant

The Principles and Practice of Electrical Epilation

Third edition

Sheila Godfrey, FIE, DRE, FBAE, FBABTAC

LONDON AND NEW YORK

First published by Butterworth-Heinemann

First published 1992
Second edition 1996
Third edition 2001

This edition published 2011 by Routledge
2 Park Square, Milton Park, Abingdon, Oxfordshire OX14 4RN
711 Third Avenue, New York, NY 10017, USA

First issued in hardback 2015

Routledge is an imprint of the Taylor & Francis Group, an informa business

Notice
No responsibility is assumed by the publisher for any injury and/or damage to persons or property as a matter of products liability, negligence or otherwise, or from any use or operation of any methods, products, instructions or ideas contained in the material herein. Because of rapid advances in the medical sciences, in particular, independent verification of diagnoses and drug dosages should be made

British Library Cataloguing in Publication Data
Godfrey, Sheila
The principles and practice of electric epilation – 3rd ed.
1. Hair – Removal
I. Title
617.4'779

Library of Congress Cataloguing in Publication Data
Godfrey, Sheila
The principles and practice of electrical epilation/Sheila Godfrey – 3rd ed.
p. cm.
Includes bibliographical references and index.
1. Hair – Removal 2. Electrolysis in surgery I. Title.
RL115.5 G64
617.4'779-dc21 2001025554

ISBN 13: 978-1-138-13386-0 (hbk)
ISBN 13: 978-0-7506-5226-1 (pbk)

Contents

Preface

The purpose of this book is to provide up-to-date information on all aspects of electro-epilation for both practising electrolysists and students studying for the following qualifications or affiliations:

- Diploma in Remedial Electrolysis
- British Association of Electrolysists
- Exdexcel
- CIBTAC (Confederation of International Beauty Therapy and Cosmetology)
- City and Guilds of London Institute (CGLI)
- VTCT Vocational Charitable Trust
- ITEC International Therapy Examination Council

Many changes have taken place in the field of electrolysis during the past thirty years. The wide range of changes that have taken place include:

- Health and Safety at Work Act 1974
- The introduction of disposable sterile needles
- Local Government Miscellaneous Provisions Act 1982
- Changes relating to technology in the development and manufacture of equipment
- AIDS/HIV
- The rising incidence of hepatitis B
- The identification of hepatitis C, D, E and G
- The introduction of the blend epilation technique into the UK
- Computerised electro-epilation machines
- The introduction of the National/Scottish Vocational Qualification Level 3
- The advent of laser and intense pulsed light systems for long-term hair removal
- Many changes in legislation with regard to employment of staff

Knowledge of allied subjects such as endocrinology, hygiene, anatomy, physiology and electricity is essential to a thorough understanding of why hair growth occurs and how this problem, which causes distress to very many people, can be treated safely and effectively.

Companies consulted

Michael Dufty Partnership – Chartered Accountants, Birmingham
Ballet Needles, Arand Ltd, Paris
Carlton Professional, Taylor Reeson Laboratories, Sussex
Continental Trades and Industry (CTI), Holland
E.A. Ellison, Coventry
Hairdressing and Beauty Equipment Centre, London
House of Famuir, Sandy, Bedfordshire
HSBC Insurance Brokers, Haywards Heath, Sussex
International Hair Route Magazine, Ontario
Lloyds TSB Bank plc
Solihull Hospital, Microbiology Department

Acknowledgements

I wish to express my thanks to the friends, colleagues and business associates who so willingly provided information for the three editions of this book.

My particular thanks go to Angela Barbagelata-Fabes, the late Rod Fabes and engineers Paul Atherton and Chris Johnson of Carlton Professional; Joseph Asch of Ballet needles, Paris for his valuable input into all three editions, particularly in relation to the provision of photographs; Romano Scavo of CTI, Holland for his help with technical information and photographs; James Paisner of Synoptic Products for information on sterilization requirements/standards in the USA; and Derek Copperthwaite of *International Hair Route Magazine* for providing material for the history of the epilation needle.

Special thanks and appreciation go to Moira Paulusz, Deputy Editor, *Health and Beauty Salon*, for her invaluable help with research and meticulous attention to detail on checking the manuscript; to Elaine Leek, the freelance copy editor, for helping me meet the publication deadline; also to Dawn Ward, Jennifer Cartwright, Janice Brown (Epilation Trainer, House of Famuir) and Bill Peberdy for their support and encouragement. Thanks are due also to Correna Lewis for her help in research and typing of the manuscript, Rita Roberts for her sympathetic teaching and support throughout the years, and last but not least, my American colleague John Fantz of California for sharing his knowledge on the history and development of blend epilation.

Photographs supplied by:

Angela Barbagelata-Fabes, Carlton Professional, Figures 13.5 (middle), 22.1 (top), 22.2 (bottom) and 22.6 (top)
Ballet Needles, Figure 9.10
Cosmetronic UK Ltd, Figure 22.4
Hairdressing and Beauty Equipment Centre, Figures 19.8 and 22.6 (middle)
John Fantz, Figures 11.1, 13.1 and 13.2
Joseph Asch, Figures 13.15, 13.16, 13.17 and 14.1
House of Famuir, Figures 22.1 (middle) and 22.7
E. A. Ellison, Figures 9.6(b), 9.8, 19.7, 22.1 (bottom), 22.2 (top), 22.3 and 22.5 (top and middle)
Continental Trades and Industry (CTI), Figures 9.6(c), 9.9, 13.5 (top), 22.5 (bottom) and 22.6 (bottom)

1 The skin

The skin is the largest organ of the body and is vital to the health and well-being of the individual. It has a number of functions that must be fully understood in order to obtain the maximum benefits without side effects when applying electrical epilation.

The skin provides shape and an outer, waterproof covering for the body. Functions include:

Sensation – touch, heat, cold, pain.
Heat regulation – blood vessels, perspiration.
Absorption – of certain medications via creams and patches such as hormone replacement therapy (HRT).
Protection – acid mantle against bacteria and micro-organisms, waterproof, against ultraviolet (UV) light, against knocks and blows.
Excretion – of toxins and waste products onto the skin's surface in perspiration.
Secretion – of sebum onto the skin's surface which combines with perspiration to form the acid mantle.

The practising electrolysist should be conversant with the anatomy and physiology of the skin, and should be able to recognize the differences between a healthy skin and one suffering from disease or disorders.

The skin consists of three main layers:

- Epidermis – the outer layer, composed of several layers of dead keratinized cells.
- Dermis – consists of two divisions: (a) the reticular layer and (b) the papillary layer.
- Subcutaneous layer – found immediately below the dermis.

Structures contained within the skin include:

- Hair follicle.
- Hair.
- Apocrine and eccrine sudoriferous (sweat) glands.
- Exocrine glands.
- Blood and lymph supply.
- Nerve endings.
- Sebaceous glands.

The thickness of the skin varies from one part of the body to another, being thickest on the soles of the feet and palms of the hands and thinnest on the lips.

The epidermis

The epidermis is the outermost layer of the skin and consists of flattened, keratinized cells that shed, or desquamate, on a regular basis. The speed of desquamation depends on a number of factors:

- Faster in young skin – slower as we age.
- Smoking.
- Exposure to the elements, such as wind, extremes of temperature, central heating.
- Skin care routine.
- Health of the individual.

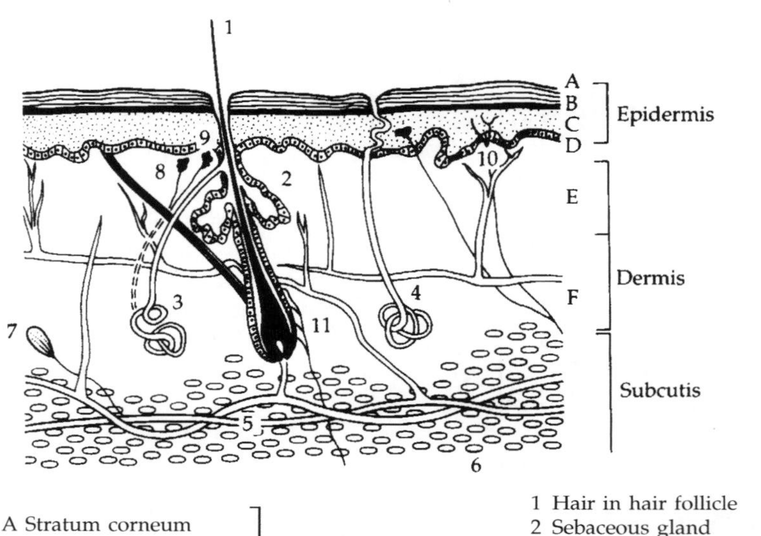

A Stratum corneum
B Stratum granulosum
C Prickle cell layer
D Basal cell layer
] Epidermis
E Papillary layer
F Reticular layer
] Dermis

1 Hair in hair follicle
2 Sebaceous gland
3 Apocrine sweat gland
4 Eccrine sweat gland
5 Blood vessels
6 Fat cells
7 Pacinian corpuscle–pressure/touch
8 Kraus bulbs–cold
9 Raffini–heat
10 Free nerve endings–pain
11 Nerve endings to follicle

Figure 1.1 Cross-section of the skin

The epidermis is non-vascular and consists of five layers:

- Stratum corneum – horny layer.
- Stratum lucidum – clear layer.
- Stratum granulosum – granular layer.
- Stratum spinosum – prickle cell layer.
- Stratum germinativum – basal cell layer.

Stratum corneum

This layer is also known as the horny layer and forms the outer layer of the skin. It consists of flattened, horny, hard, keratinized cells without a nucleus. These outer cells are constantly being shed and replaced from the lower layers. Air gaps found between cells in this layer make it a poor conductor of heat and light.

Stratum lucidum

This layer, which is present only in the palms of the hand and soles of the feet, consists of flattened, closely packed cells. Traces of flattened nuclei

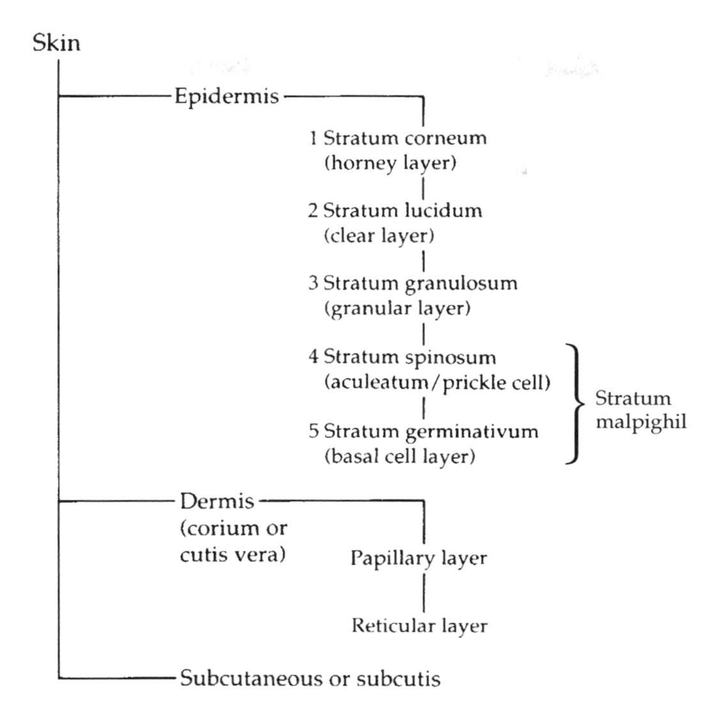

Figure 1.2 Layers of the skin

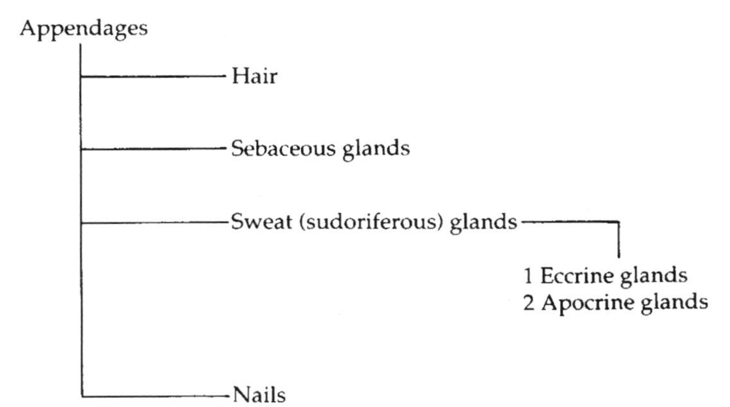

Figure 1.3 Appendages of the skin

may be found. The cells are transparent, which allows the passage of light to the deeper layers.

Stratum granulosum

This is the third layer of the epidermis, and consists of granular cells. The granules contain keratohyaline, which is an intermediate substance in the formation of keratin. Nuclei and other cell contents begin to disappear. (The granules consist of a substance called eleidin which is the intermediate substance in the formation of horn: *Gray's Anatomy*, page 1138.)

Stratum spinosum

This layer is often referred to as the prickle cell layer (or the stratum aculeatum) and is situated immediately under the stratum granulosum. The cells are covered with numerous fibrils that connect the surfaces of the cells.

These cells are known as prickle cells. Between the cells are fine intercellular clefts that allow the passage of lymph corpuscles. Pigment granules may be found here. Cells have flattened slightly.

Stratum germinativum

This is the deepest layer of the epidermis and is known as the basal cell layer. This layer, together with the stratum spinosum, is known as the *Malpighian layer* or *stratum malpighii*. The germinativum is closely moulded onto the papillary layer of the dermis below. The cells are keratinocytes, contain nuclei and are capable of cell division. They are larger than cells in the upper layers and are columnar in shape. The contents are soft, opaque and granular. Cells contain melanocytes which are responsible for producing *melanin*. The basal cell layer surrounds the hair shaft and bulb.

The dermis

The dermis, also known as the *corium* or *cutis vera,* is situated immediately below the epidermis. Blood vessels, lymph vessels and nerve endings together with collagen and elastin are contained within the dermis. The dermis forms the bulk of the skin and consists of two parts: the papillary and reticular layers.

The papillary layer

This is situated on the free surface of the reticular layer and consists of a number of small, highly sensitive projections known as papillae. The papillae are composed of very small, closely interlaced bundles of fibrillated tissue. Within this tissue is a capillary loop carrying oxyhaemoglobin to other tissues within the dermis. Tactile corpuscles can be found in some of these papillae.

The reticular layer

This layer forms the bulk of the dermis and is found immediately below the papillary layer. The majority of the collagen and elastin fibres are found in this layer. Fibroblasts are spindle-shaped cells which are responsible for the production of collagen and elastin. Mast cells are also found here and are responsible for the release of heparin and histamine. The function of histamine is to dilate bloods vessels, and the function of heparin is to prevent blood from clotting. The typical mast cell is large, rounded or spindle-shaped with one, or occasionally two, nuclei.

The dermis is composed of connective tissue that contains a ground substance or matrix containing most of the skin's water content in which are suspended:

- Cells – fibroblasts and mast cells.
- Fibres – collagen, elastin and reticulin.

There are three different connective tissues found in the dermis. These are:

- White fibrous tissue/collagen.
- Yellow elastic tissue/elastin.
- Reticulin.

Collagen fibres are also known as *white fibrous tissue* and form 75% of the total connective tissue in the dermis. These fibres are embedded in a ground substance of colloidal gel. Fibroblasts are interspersed between bundles of collagen. Collagen gives the skin its toughness and resilience.

Elastin, or *yellow fibres*, forms only 4% of the connective tissue. These fibres run parallel, or obliquely to the collagen and enclose the bundles. Elastin gives the skin its elasticity.

Reticulin fibres ensure the stability between the dermis and epidermis.

Structures found within the dermis are:

- Sebaceous glands.
- Sudoriferous (sweat) glands – apocrine and eccrine.
- Arrector pili muscles.
- Hair follicles.

Sebaceous glands

Sebaceous glands are situated within the dermis with their ducts opening into the hair follicle. Occasionally some may open directly onto the skin's surface. They are absent from the palms of the hands and soles of the feet. Each gland is constructed of a single duct that ends in a cluster of secretory saccules similar to a bunch of grapes in appearance.

Sebaceous glands are highly sensitive to androgens, which stimulate the growth of the gland and the production of sebum. Sebum production is increased at puberty and decreases with age. These glands secrete sebum via the hair follicle onto the skin surface. The purpose of sebum is to keep the hair pliable and to lubricate the skin. Sebum is also responsible for making the skin waterproof and plays a major role in the formation and maintenance of the 'acid mantle'.

Sebum contains fatty acids, esters and other substances.

Sudoriferous glands

Sudoriferous glands, or **sweat glands,** are found all over the skin surface. Their function is to regulate the skin temperature through the evaporation of sweat on the skin's surface and also to excrete a small amount of waste products. These glands can be divided into eccrine and apocrine glands.

Eccrine glands are present in large numbers – between two and five million in total. They are found in all parts of the skin with the excep-

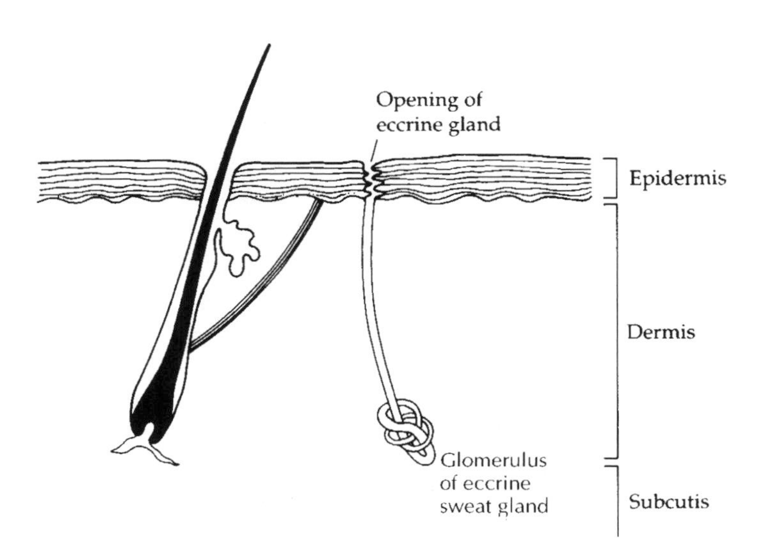

Figure 1.4 Eccrine sweat gland

tion of the mucous membranes. These glands are independent of hair follicles and open directly onto the skin's surface. The eccrine gland is a down-growth of the epidermis that reaches into the deeper layers of the dermis where it forms a coiled ball known as the glomerulus. The duct is straight in the upper dermis, and becomes twisted, like a corkscrew, in the epidermis.

The secretion from the eccrine glands is a clear, watery fluid consisting of 99–99.5% water, together with some chlorides, lactic acid and urea.

Apocrine glands are larger than eccrine glands (see Figure 1.5). Their ducts open directly into the hair follicle – only occasionally do they open directly onto the skin surface. They are coiled, tubular glands with a duct leading down to a coil of secretory tubules. These glands are under hormonal control and become active at puberty. Stimulation is brought about by stress, fright, pain or sexual activity. They can be found in the axilla, groin and around the nipples.

Secretion from these glands is a sterile, whitish fluid which contains proteins, carbohydrates and other substances.

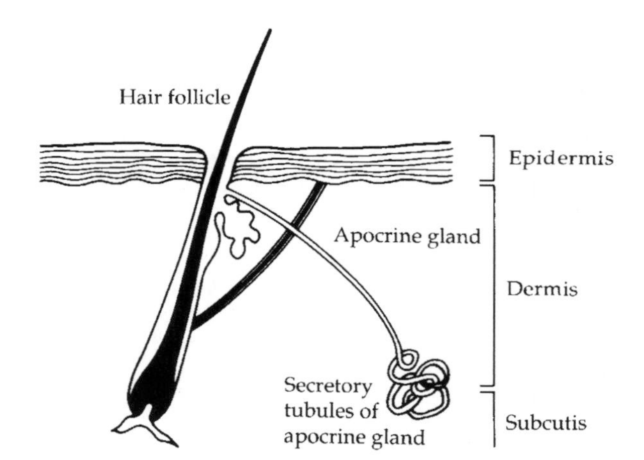

Figure 1.5 Apocrine gland

Functions of the skin

The skin performs a number of functions, as briefly mentioned on page 1.

Secretion

The sebaceous glands are responsible for the production of sebum and its secretion onto the skin's surface via the hair follicle. The role of sebum is to mix with perspiration to form the acid mantle, so maintaining the pH balance. Sebum also provides a waterproof film on the epidermis and helps to lubricate hair.

Heat regulation

Heat is regulated in the following manner. When the body or skin temperature is too high the blood vessels dilate to allow the loss of heat from the blood vessels on to the skin. This, in combination with the evaporation of sweat, results in a reduction of heat. When the body is cold, heat is conserved

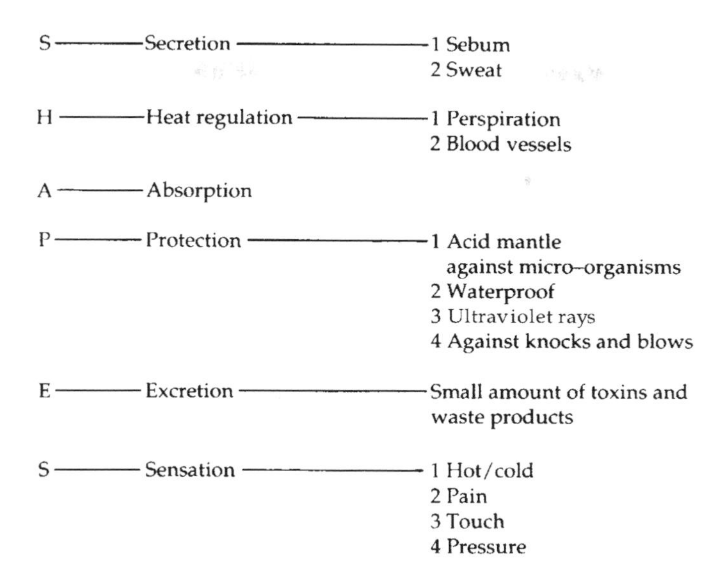

Figure 1.6 Functions of the skin

internally by the vaso-constriction of blood vessels that carry the blood away from the skin and a reduction in the amount of perspiration secreted by the sudoriferous (sweat) glands, so keeping the heat within the body.

Absorption

The administration of certain medications such as hormone replacement therapy by means of impregnated patches applied to the skin shows that the skin is capable of absorbing some substances.

Protection

The skin protects the body in several ways:

- Melanocytes, found in the epidermis, produce melanin which helps to protect the skin from the harmful rays of UV (ultraviolet) light.
- The 'acid mantle' refers to a fine film of sebum and sweat found on the surface of the skin. The purpose of the acid mantle is to inhibit the growth of bacteria and prevent the entry of micro-organisms into the skin. The pH scale is measured on a scale of 0–14, with 7 being neutral. The lower the number, the higher the acidity. The normal pH of the white skin is between 4.5 and 6. The acid mantle is often destroyed through over enthusiastic use of soaps and harsh detergents.
- The subcutaneous layer acts as a buffer, giving protection from knocks and blows.
- The keratinized cells in the epidermis prevent the entry of harsh chemicals, bacteria and viruses.

Figure 1.7 The pH scale

0	1	2	3	4	5	6	7	8	9	10	11	12	13	14
Acid							N	Alkaline						

Excretion

Small amounts of waste products and toxins are eliminated from the body onto the skin's surface via perspiration. The amount of waste matter increases during times of stress and ill health.

Sensation

The skin is the principal seat of sensation. Due to the abundant supply of sensory nerves contained within the papillary layer of the dermis, it is possible for the skin to respond to heat, cold, pain and pressure. The nerve supply also surrounds the hair follicle, arrector pili muscle and sweat glands.

Nerves within the skin end either in a corpuscle or free endings. It is the free-ending nerves that are responsible for the sense of pain. Nerves ending in corpuscles are of three types:

- Pacinnian – responsible for touch or pressure.
- Krause bulbs – responsible for sensation of cold.
- Organs of Raffini – responsible for feeling heat.

Meissner corpuscles are unique sensory receptors found only in the hands, feet and digits.

Formation of vitamin D

Vitamin D is formed by the action of ultraviolet light on dehydrocholesterol in the skin. Vitamin D aids in the formation and maintenance of healthy bone.

Natural moisturizing factor

The function of the natural moisturizing factor is to aid in the prevention of loss of moisture from the epidermis. As the cells within the skin structures move up towards the surface, moisture is pressed out of the cell which results in each cell being coated with a sticky intercellular glue, thereby holding the stratum corneum together.

Black skins

There are a number of differences between the structure of black and white skin, which should be considered when giving electrical epilation.

Sudoriferous glands

In black races the sudoriferous glands are larger and more numerous. The glomerulus is rounder but not as coiled. The secretory duct, which opens onto the surface of the skin, is longer and more noticeable.

There are a larger number of eccrine glands per square cm, with the pH balance being between 3.5 and 5.6.

In black races melanin pigment is found in the glomerulus and sometimes in the wall of the excretory duct.

Sebaceous glands

The sebaceous glands are not only larger and more numerous, but approximately 10% open directly onto the skin's surface, with the other 90% opening into the hair follicle.

Pigmentation

It is important to remember when treating black skin with electrical epilation that it is not easy to detect erythema. Therefore care must be taken to ensure that the skin does not become over treated due to too much heat being applied in any one area.

Stratum corneum

In a black skin this is thicker than in a white skin, with black skin desquamating more easily. Pigmentary granules are present in the desquamating cells of black skin but not of white (H. Pierantoni, *Essential Notions of Black Skin*, 1977).

Keloids

These are more frequently found on black skins. A keloid is an excessive formation of scar tissue at the site of an injury to the skin. It is essential that the electrolysist discusses this possibility during the consultation before giving any treatment. Keloids can continue to grow and harden for several months after the initial injury.

Asian skin

Asian skin is normally finer in texture than black skin. There is a wide variation in colour from light brown to dark brown. In many instances hair growth tends to be dense but very fine and dark. However, growth may be coarse and dark, lying deeper in the skin than the fine, downy hairs. Fine hairs usually grow from follicles that are inclined to be small, straight and lie close to the skin's surface.

The skin is prone to sensitivity during treatment so care must be taken to avoid over-treating any one area. Excess application of heat to the skin will lead to increased pigmentation at the treatment site. This may not be evident immediately but take several days (or in some instances weeks) to develop. This pigmentation will take several months if not longer to fade. Asian skin will scar easily if the electrolysist takes insufficient care during treatment.

Oriental skin

Oriental skin is prone to pigmentation, pit marks and discoloration due to exposure to heat and UV (ultraviolet) irradiation.

Pigment variations between different skin colours

Several substances affect the colour of the skin; these vary between different races:

- Melanin.
- Melanoid.
- Carotene.
- Haemoglobin.
- Oxyhaemoglobin.

Summary

- Skin consists of three main layers: (1) epidermis, (2) dermis and (3) subcutaneous layer.
- The epidermis is the outer layer and varies in thickness from one area to another.
- The epidermis does not contain blood vessels, sheds the outer keratinized cells regularly, and also acts as a protective barrier.
- Cells that contain melanocytes that produce melanin are found in the basal cell layer and surround the hair shaft and bulb.
- The dermis lies immediately under the epidermis. It contains blood and lymph vessels, nerve endings, sebaceous glands, sweat glands, collagen, elastin, fibroblasts, hairs and hair follicles.
- Functions of the skin: sensation, heat regulation, absorption, protection, excretion and secretion.
- Colour of the skin is affected by: melanin, melanoid, carotene, haemoglobin and oxyhaemoglobin.
- Black skin contains a higher proportion of pigment granules than white skin; the epidermis is thicker in black skin with a tendency to develop keloid scars.

- Asian skin is sensitive to heat and prone to hyperpigmentation if excessive heat is used during treatment. Follicles are usually fine, small and straight with dense, fine, dark hair growth.

Review questions

1. Name and describe the layers of the epidermis.
2. Name and describe the layers of the dermis.
3. Name and describe the functions of the skin.
4. List the structures contained within the dermis.
5. Explain the difference between an eccrine gland and an apocrine gland.
6. List the substances that affect the colour of the skin.
7. What is a keloid?
8. State the differences in structure between black skin and white skin.
9. What causes hyperpigmentation in Asian skin?
10. Give the location and function of the sebaceous glands.
11. What is meant by the term 'acid mantle' of the skin?
12. What is the normal pH of the skin?
13. State the importance of the correct pH balance.
14. List the substances that affect the colour of the skin.

2 Dermatology

Electrolysists will find during the course of their professional life that knowledge of the skin's structure and function is not enough. The skin is the largest organ of the body and in many instances reflects the health of the individual. The electrolysist should therefore:

1 be able to recognize the different skin lesions;
2 be able to identify the most common skin diseases;
3 know the causes of these diseases;
4 be able to recognize when a disease is contra-indicated to electro-epilation;
5 know when to refer the client for further investigation by the medical profession.

Many clients expect the electrolysist to be the font of all knowledge when it comes to recognizing skin disorders and diseases. The same clients also expect an instant remedy, which is not always possible. It is essential that the electrolysist should be aware of when to treat and when to refer on to the medical profession when a client appears in the treatment room with a skin disease.

Examination of the skin should include the following:

- *Site(s) and distribution:* e.g. psoriasis usually appears on the knees, elbows and scalp. Rosacea appears on the cheeks and nose in a butterfly pattern. Acne vulgaris appears on the face, back and/or chest.
- *The types of lesions present*, such as milia, comedones, silvery scales, pustules, papules to name but a few.
- *Size, shape, outline and border*: e.g. is the lesion symmetric (can be divided into two equal halves) or irregular in shape? Is the surface smooth or rough? Is the border well defined or indistinct?
- The *colour of lesions* is a useful guide to indicating the possible skin disease/disorder, e.g. red, white, brown, brown/black, silvery scales.

Many skin conditions present similar lesions. Thorough examination in a good light is essential, as is careful questioning, which should include family background. In a number of instances accurate diagnosis is not possible without referral to the medical profession for further investigation. Guttate psoriasis and pityriasis rosea show many similarities and can easily be confused. In the early stages of pityriasis rosea the herald patch may be mistaken for ringworm at the initial inspection.

The health and condition of the skin is affected by a number of internal and external influences:

- *Hormonal:* low oestrogen levels cause the skin to become dry and lose collagen content. Androgens cause the skin to become oily. Steroid creams used over a prolonged period of time may result in thinning of the skin structures.

- *Allergies* may be caused by food intolerances, alcohol, drugs such as aspirin and penicillin, cosmetic preparations, perfume, lanolin, preservatives and certain chemicals such as hair dyes.
- *Environment:* poor hygiene, exposure to ultraviolet light, humidity which affects pH balance, and central heating which may cause excessive dryness.
- *Genetic predisposition:* e.g. eczema.
- *Nutritional deficiencies* resulting in lack of essential vitamins and minerals: e.g. low levels of vitamin A leave the skin dry, insufficient vitamin C affects the strength of capillary walls, which means that the skin bruises easily.

Stress over a prolonged period of time also has a detrimental effect on the skin; fine lines or dehydration may develop. Excess sebum may be produced due to increased levels of androgens, which in turn may also trigger increased hair growth.

Common skin lesions

Skin lesions show alterations in the appearance of the skin. Depending on the severity, lesions are classified as primary, secondary or tertiary. The terms used to describe changes in the skin are shown in Table 2.1.

Changes in skin colour and appearance

- *Hyperkeratosis*: refers to excess keratinization of cells, or over-growth of horny cells, e.g. scales and plaques found in psoriasis.
- *Erythema*: general redness of the skin due to temporary or permanent vasodilatation.
- *Hyperpigmentation*: describes an excess or increase in skin pigmentation, e.g. chloasma, lentigo.
- *Hypopigmentation*: describes loss of pigmentation, e.g. vitiligo.

Causes

Skin diseases and disorders may occur as a result of one of the following:

1. Bacterial infection.
2. Viral infection.
3. Fungal infection.
4. Parasites.
5. Exposure to allergens.
6. Hereditary/genetic predisposition.
7. Pigmentation abnormalities.
8. Hormonal influences.
9. Stress related.
10. Carcinogenic/cancer.
11. Light-induced or photosensitive.

Bacterial infections

Boils

Boils are caused by a staphylococcal infection of the hair follicle. The area becomes red, swollen and painful. Heat and oedema are present. After a short period of time the centre fills with pus and eventually bursts. A scar frequently develops after the boil has healed.

Common sites for the development of boils are the axilla, the back of the neck, the buttocks and thighs. Their appearance is a sign of lower resistance to infection due to being tired and run down.

Table 2.1

Macule	Flat, small patch of increased pigmentation or discoloration, e.g. a freckle
Papule	Small, raised elevation on the skin, less than 1 cm in diameter, which may be red in colour, e.g. acne and rosacea
Pustule	Small, raised elevation of the skin which contains pus, seen in acne and rosacea
Nodule	A well-defined, solid lump more than 1 cm in diameter, seen in boils and rodent ulcers
Plaque	A well-defined, disc-shaped elevated area of skin, seen in psoriasis
Vesicle	Small lesion (less than 0.5 cm in diameter) containing fluid, i.e. small blisters seen in herpes simplex, herpes zoster and impetigo
Bulla	Large lesion, more than 2–3 cm in diameter, containing fluid, i.e. blister
Comedone	May be closed or open. This lesion is a collection of sebum, keratinized cells and certain waste substances that accumulate in the entrance of the hair follicle. An open comedone is a blackhead contained within the follicle, whereas a closed comedone, or whitehead, is trapped underneath the skin's surface. Closed comedones do have a small opening on the surface of the skin
Milia	Small accumulation of fatty substances under the skin, often seen around the eyes and on the cheeks of dry skin
Ulcer	A loss of epidermis (frequently with loss of underlying dermis and subcutis). Seen in a rodent ulcer
Scale	Visible flakes of skin on the surface of the epidermis, seen in psoriasis
Crust	Also referred to as a scab, this is an accumulation of dried fluid, serum or pus on the surface of the skin, seen in impetigo
Fissure	Crack in the skin's surface, e.g. chapped lips, which can be painful
Excoriation	A secondary superficial ulceration which is due to scratching
Wheal	Well-defined raised area of cutaneous oedema, white in the centre with a red edge, seen in urticaria
Scar	Appears during the healing process. Skin tissue may be smooth and shiny or form a depression in the surface, e.g. an ice-pick scar seen in acne
Telangiectasia	Persistent vasodilation of the capillaries in the skin
Keloid	Over-growth of scar tissue with a raised, shiny appearance
Lichenification	Areas of increased epidermal thickness with accentuation of skin

Carbuncles develop when a number of boils appear in close proximity to one another.

Erythrasma

Erythrasma is a bacterial infection produced by a Gram-positive bacillus, *Corynebacterium inutissimum*. Common sites for the development of erythrasma are the axilla, groin and sub-mammary regions, but the commonest site colonized by these bacteria is in between the toes.

The characteristics of erythrasma are the appearance of marginated brown areas with a fine, branny surface scale. Symptoms can be aggravated by increased temperature, perspiration and scratching of the area due to irritation.

Impetigo

Impetigo is a superficial, contagious, inflammatory disease caused by streptococcal and staphylococcal bacteria. It is commonly seen on the face and around the ears. Weeping vesicles dry to form honey-coloured crusts. The bacteria can be transmitted by dirty fingernails and towels. Impetigo may also arise as a secondary infection from an existing condition such as scabies.

Folliculitis

Folliculitis is characterised by pustules and inflammation, arising in and around the upper part of the hair follicle, as a result of staphylococcal infection. Folliculitis is to be found in hairy areas, in particular the male beard, which is often aggravated by shaving.

Viral infections

Conditions caused by a viral infection include warts, herpes simplex and herpes zoster, all of which are contagious.

Warts

Warts appear in several forms. They are well-defined, self-limiting, benign tumours, which vary in size and shape. Warts are caused by infection with the human papilloma virus. The *common wart* is skin coloured or brownish with a smooth or rough surface, which varies in size from a pinhead to the size of a pea. These warts are usually found on the fingers and hands, elbows, knees and sites of minor trauma. *Plane warts* are found on the face, forehead, back of hands or front of knees. They are smooth in texture with a flat top and frequently brownish in colour. They are most commonly found in children. The *plantar wart* is the size of a pea, or a little larger. These are found on the sole of the foot and may be very painful. This type of wart is also called a verruca. The *filiform wart* hangs down from the skin's surface and may grow up to 6 mm in length. These are quite thick in diameter. The face and neck are the most usual sites for this type of wart. Filiform warts are hard and keratinous.

Herpes simplex

Herpes simplex (also referred to as cold sores) is normally found on the face and around the lips. The onset is quite rapid, beginning with an itching sensation, shortly followed by erythema and groups of vesicles. Crusts form where the vesicles weep. The condition clears up in approximately two to three weeks, but will reappear in the same area in times of stress, ill health or exposure to sunlight. Recurrent attacks of herpes simplex begin with a tingling sensation, followed by tender, often painful blisters, which usually appear on the upper lip.

Herpes zoster

Herpes zoster is also known as shingles. The cause is the chicken-pox virus, *Herpesvirus varicella* The condition is very painful due to acute inflammation of one or more peripheral nerves. The pain may persist for up to 18 months. The lesions resemble herpes simplex, with erythema and vesicles along the line of a nerve. Areas affected include the chest and back and along the trigeminal nerve of the face.

Fungal infections

Fungal infections include tinea pedis, tinea capitis, tinea corporis and pityriasis versicolor, also known as tinea pityriasis. Tinea or ringworm is caused by a superficial fungus, which lives on the skin and feeds off dead horny cells. The fungi digest keratin. There are several forms of fungal infection, but the three which are of most interest to the electrolysist are tinea corporis, affecting the body, tinea capitis, affecting the scalp, and tinea pedis – athlete's foot – affecting the feet.

Tinea corporis

Also known as ringworm of the body, this exhibits lesions, which begin as small red papules that gradually increase in size to form a ring. These lesions, which vary in shape from round to oval, gradually clear in the centre as they increase in size. However some lesions do not clear from the centre but appear as red, scaly plaques.

Tinea capitus

This particular fungal infection affects both the hair and the scalp. Lesions appear as small oval patches on the scalp. The hairs break off near to the skin's surface, which has a scaly base.

Tinea pedis

Tinea pedis, more commonly referred to as athlete's foot, is highly contagious and can easily be acquired from damp places such as swimming pools, showers or saunas. A sign of athlete's foot is the appearance of flaking skin between the toes, which becomes soft and soggy. The skin may also split and the condition is sometimes uncomfortable. Occasionally the soles of the feet are affected.

Pityriasis versicolor (tinea versicolor)

Pityriasis versicolor is a superficial fungal infection, which affects the back, chest and axilla. This condition develops gradually and can be recognized by the appearance of well-defined fawn or coffee-coloured lesions with fine branny scales, on the neck, shoulders, upper trunk and upper arms.

Infestations by insects

Scabies

Scabies, caused by the *Sarcoptes scabiei* parasite, is a contagious parasitic infection caused by the itch mite. The female mite (arcarus) burrows into the horny layer of the skin where she lays her eggs. The eggs hatch after three to four days, and the larvae gravitate from their burrows into the adjacent hair follicles. After approximately 17 days the adult mites emerge and the whole cycle starts again. The adult female mite may live in her burrow for 6–8 weeks.

A characteristic lesion in scabies is the burrow, a white or greyish zigzag line, which may be slightly curved and scaly, which varies in length between 0.5 cm and 1 cm. Vesicles may be visible at one end of the lesion. The onset of this condition is gradual, the first noticeable signs being severe itching, which is usually worse at night. A generalized rash may then appear, with irritation becoming noticeable during the day. Papules, pustules, excoriations and crusted lesions may also develop. Common sites for this infestation are the ulnar borders of the wrists, palms of the hands and between the fingers. Other sites that may be affected are the axillary folds, the buttocks, and breasts in the female and external genitalia in the male.

Pediculosis (lice)

Pediculosis is also a contagious parasitic infestation, where the lice live off blood sucked from the skin. The female louse lays numerous eggs during her one-month life span.

Nits are small, white oval eggs stuck firmly to the hair by its capsule. They may be moved up or down a hair but cannot be removed sideways. After 6–10 days these nits hatch into larvae, developing into fully-grown lice within 1–2 weeks.

Head lice (*Pediculosis capitis*) are frequently seen in very young children who become infested at school. The condition spreads very quickly and can only be controlled by thorough treatment. The lice make no distinction between class or race and are happiest in clean hair. If not dealt with swiftly this condition may lead to secondary infection as a result of scratching, e.g. impetigo.

Body lice (*Pediculosis corporis*) are rarely seen today. They usually occur on a person with poor personal hygiene. They live and reproduce in the seams and fibres of clothing, leaving only to feed from the skin. Lesions may appear as papules, scabs, and in severe cases pigmented, dry scaly skin. Secondary bacterial infection is often present.

Eczema/dermatitis

The terms eczema and dermatitis are synonymous, in other words both terms may be used to describe the same condition. Eczema is derived from the Greek term *ekzein*, meaning to break out or boil over; dermatitis means inflammation of the derma or skin. Both terms relate to a condition that varies from a mild to a chronic inflammatory state. The term 'dermatitis' is preferred by most dermatologists.

Dermatitis may be due to a genetic predisposition, or to internal or external influences. When genetic predisposition is the root cause it is not unusual to find a history of asthma and/or hay fever in the family.

Clinical signs of dermatitis begin with small, itchy patches of erythema, which may gradually increase in size. Oedema, fissures (cracks), scales and hyperkeratosis are other symptoms associated with this condition. In severe cases the skin weeps where fissures are present or where the surface has been scratched.

Acute dermatitis

Acute dermatitis may follow a single exposure to a chemical or irritant. The affected area becomes red (erythema), swollen and may itch. In some instances the inflammation will subside and the skin will return to normal after a short period of time. However, the inflammation may increase with the development of vesicles, being followed by the formation of crusts.

Chronic dermatitis

Skin that is affected by dermatitis for several weeks tends to thicken and develop pigmentation. Scratching of the area tends to aggravate the condition.

Contact dermatitis/eczema

This is caused by a primary irritant which causes a reaction in susceptible individuals. The reaction may occur after a short exposure to an irritant or may build up over a period of time after repeated contact.

Substances that cause this type of dermatitis include: acids; alkalis; solvents; cosmetic preparations, in particular perfume and lanolin; detergents; nickel, e.g. ear-rings and suspenders; household polishes, plus certain house and garden plants, e.g. primulas, tulips, chrysanthemums and celery. Lesions are normally localized to the area of contact.

Phototoxic dermatitis

This arises after exposure of a phototoxic substance on the skin to ultraviolet light (sunlight or sunbeds). Phototoxic dermatitis is similar in appearance to sunburn. Pigmentation of the exposed area often occurs. Citrus-based perfumes and essential oils on the skin, certain herbal remedies such as St John's Wort, and some medications such as Prozac are very photosensitive.

Allergic dermatitis/eczema

This is generally more widespread and does not appear immediately after the first exposure to the irritant or allergen. The individual gradually builds up sensitivity to substances such as perfume or certain dairy products. Cow's milk is a common cause of dermatitis/eczema in children. Once sensitivity to an allergen has developed, further contact, even after a period of weeks or months, will result in the recurrence of dermatitis.

The sites most commonly affected are the hands and feet, but any area of the body may react when exposed to the offending substance. As with contact dermatitis, erythema is present and the skin becomes itchy with a build-up of scales. In chronic cases small blisters, hyperkeratosis and fissures develop. Scratching will aggravate the condition.

Seborrhoeic dermatitis (seborrhoeic eczema)

Seborrhoeic dermatitis is a mild to chronic inflammatory disease of hairy areas well supplied with sebaceous glands. An increase in sebum production with an alteration in chemical composition may or may not be present. Common sites for this condition are the scalp, the face, axilla, submammary folds and the groin.

The skin may appear to have a grey tinge or be dirty yellow in colour. The onset of the condition is gradual. Clinical signs may show slight redness and scaling of the naso-labial folds, dandruff in the eyebrows and possibly deep-seated pustules affecting the follicles in the beard area of the adult male. When the scalp is affected greasy scales or dry, scaly plaques will be seen.

Psoriasis

Psoriasis is a chronic inflammatory condition of the skin. Although the cause is not known, there is no doubt that a genetic factor exists. Any age group can be affected, but it very rarely appears in children under five years of age. Psoriasis is aggravated by stress, bacterial throat infection and trauma to the skin, but is improved by exposure to sunlight.

This disorder can be recognized by the development of well-defined red plaques which vary in size and shape, covered by white or silvery scales.

When the scales are removed the surface underneath will be smooth and red and will show pinpoint bleeding. The edges of plaques are well defined.

Any area of the body may develop psoriasis, but the most commonly affected sites are the extensor surfaces, chest, abdomen, face, elbows, knees and nails.

Pityriasis rosea

Pityriasis rosea begins with the appearance of the herald patch 7–10 days prior to development of other lesions. The herald patch usually appears on the trunk. This initial lesion presents itself as a scaly patch with a slightly raised edge, which clears in the centre. At this stage it is possible to confuse pityriasis rosea with tinea corporis. The disease has no known cause. It can be termed self-limiting, and usually runs its course within 6–8 weeks. Pityriasis rosea is most commonly seen in children and young adults.

Clinical signs are oval-shaped lesions with well-defined edges and a scaly surface. Macules and papules may also be present. Pityriasis rosea appears mainly on the trunk and very rarely affects the face, hands or feet.

Acne vulgaris

Acne vulgaris is a condition that causes much stress and embarrassment in both sexes. Onset is gradual, appearing most frequently at puberty and persisting for some considerable time. Acne vulgaris rarely lasts beyond the age of 30. Salon treatments, whilst not making any claims to cure the condition, do help to keep lesions under control and scarring to a minimum. Exposure to ultraviolet light and sunbed, in some instances, can help to improve the condition. The condition is often aggravated during the week before menstruation, during times of stress, e.g. examinations, interviews, important occasions, and in humid climates. Areas most affected are the face, neck, chest and back.

Acne vulgaris is due to a defect of the sebaceous glands that leads to over-production of sebum. It is primarily androgen-induced and may indicate hypersensitivity of the sebaceous glands to circulating hormones.

Clinical features are the presence of comedones – both open and closed – papules, pustules, cysts, scars and hyperkeratosis of the horny layer. The skin is usually oily due to excess sebum on the epidermis. Ice-pick or keloid scars may also develop in susceptible individuals.

The main factors that contribute to the development of acne lesions are:

1. excess sebum production from the sebaceous glands;
2. obstruction of the follicle opening by keratinized cells;
3. inflammation as a result of leakage of contents from the pilosebaceous canal into the surrounding dermis;
4. excessive infection of the pilosebaceous ducts with bacteria.

Sebaceous cysts

Sebaceous cysts are round, nodular lesions with a smooth, shiny surface, which develop from a sebaceous gland. They are usually found on the face, neck, scalp and back. The cause is unknown. They are situated in the dermis and vary in size from 5 mm to 50 mm. The lesion is surrounded by fibrous connective tissue. Cysts contain masses of disintegrating epithelial cells, the contents of which are soft and cheesy.

Rosacea

The cause of rosacea is not known but it usually affects adults of both sexes from the age of 30 onwards, although it is more frequently seen from the age of 45 years. Aggravating factors are heat, hot spicy foods, hot drinks, alcohol, emotional stress, menopause, cold winds and sunlight.

The onset is gradual and begins with flushing of the cheeks and nose; telangiectasia become noticeable. The condition may then spread to the centre forehead and chin. As the condition progresses, papules, pustules and scales develop. In advanced cases, rhinophyma may occur, the characteristic signs being hypertrophy of the sebaceous glands and thickening of the skin of the nose.

Table 2.2 Comparison between acne vulgaris and rosacea

Acne vulgaris	*Rosacea*
Develops around the age of puberty, rarely occurring after the age of 30 years	Rarely develops before the age of 30 years
Comedones are usually present	Comedones do not occur
Sites normally affected: the sides of the face, chin, temples, sides of cheeks, shoulders and front of chest, back	Sites affected: the nose, centre cheeks, centre of forehead and centre chin

Urticaria

Urticaria is a condition in which the lesions usually appear rapidly and disappear within minutes or gradually over a number of hours. Other lesions may occur in the same or in other areas later on. The clinical signs of urticaria are the development of red wheals, which may later become white. The area becomes itchy or may sting, e.g. after touching stinging nettles.

There are numerous causes for urticaria, some of which are an allergic reaction to certain foods, e.g. strawberries and shell fish, drugs such as penicillin or aspirin; inhalants such as house dust, animal fur and pollens. Other causes include stress, dermagraphic skin, sensitivity to light, heat or cold.

Pigmentation disorders

Some conditions are due either to excess pigmentation or to lack of pigmentation in the skin. These include vitiligo, lentigo and chloasma. Generalized pigmentation may be associated with a systemic disease such as pituitary tumour or Addison's disease.

Vitiligo is the name used to describe lack of pigmentation in the skin. Any area can be affected and the size of patches, which may be oval or irregular in shape, can vary from quite small to covering extensive areas. It is most commonly seen on the face and hands. Both sexes of any age group can be affected. The cause is unknown. Patches lacking pigmentation are very sensitive to sunlight, and burn easily.

Lentigo is the technical term given to freckles. They tend to appear after exposure to sunlight, and fade during the winter months. They are most commonly seen on red-haired and fair-skinned people.

Chloasma is patches of increased pigmentation usually seen on the face during pregnancy. The oral contraceptive pill may also give rise to chloasma. This type of pigmentation usually fades after the pregnancy has ended. It may also occur during the menopause.

Hyperpigmentation is due to sensitization of the skin on exposure to sunlight. Substances such as perfume, certain cosmetic preparations, and citrus-based essential oils, e.g. bergamot, all react with sunlight to produce increased pigmentation. This type of pigmentation, which is known as Berloque dermatitis, may take years to fade.

Naevi and benign tumours

There are a number of different naevi, some of which can easily be identified while others cannot. Some naevi may respond to cauterization, some are best left alone and others benefit from medical treatment. **With any naevus that may be suitable for cauterization, electrolysists should not take it upon themselves to make a final decision but should refer clients back to their general practitioner for written agreement prior to treatment.**

Pigmented naevi

These may also be referred to as moles. They appear as flat or raised, round, smooth lumps on the surface of the skin. They vary in size and in colour, from pink to brown or black. It is very rare for this type of lesion to become malignant. They are caused by changes in the melanocytes which give rise to cells called 'naevus cells'.

Port-wine stain

This is also known as deep capillary naevus, or mature haemangiomata, and is due to an accumulation of dilated capillaries in the dermis and deeper layers of the skin. It is present at birth and may vary in colour from a pale pink to deep purple. It has an irregular shape and is not raised above the skin's surface. These naevi are often found on the face but may also appear on other areas of the body.

Haemangiomatoma

These appear as a small, raised, red papule which bleeds easily. The cause is unknown but they often appear after vigorous squeezing of skin lesions such as comedones or papules. They are easily removed by short-wave diathermy.

Spider naevus

This refers to a collection of telangiectasia, which radiate from a central papule. They often appear during pregnancy and if left alone may disappear on their own. Alternatively they are easily dealt with by short-wave diathermy. When multiple naevi are present a liver disease may be indicated.

Keloid

This is the term used for an overgrowth of scar tissue. The surface may be smooth and shiny, or ridged. Keloid scars usually appear over the seat of a previous lesion or after surgery along the site of the incision. The onset is gradual and is due to an accumulation or increase of collagen in the immediate area. The colour varies from red, fading to pink or white. There is a racial predisposition to the development of keloids, and they are very common on Negroid skin.

Fibroma

These are hard, painless nodules, which may be shiny or firm. They are normally found on the extremities.

Seborrhoeic warts

Also referred to as senile warts, these are not caused by a viral infection and are not related in any way to age. They are normally found on the trunk but may also appear on the face or other areas. Clinical signs indicate hyperkeratosis with increased pigmentation, which varies from one lesion to another. The surface is rough and uneven.

Malignant tumours

Malignant melanoma

This is a rare tumour that develops from a pigmented naevi. Its main characteristic is a blue–black nodule that increases in size, shape and colour. These tumours are most commonly found on the head, neck and trunk. Development of this tumour may be either slow or rapid. It should be emphasized that when a mole or tumour shows any change in shape, size or colour, or begins to itch, a doctor should be consulted.

Basal cell carcinoma

This is also referred to as a basal cell epithelioma or more commonly a rodent ulcer. Clinical characteristics of a rodent ulcer are a raised, nodular, shiny lesion with a pearly edge. The surface may become ulcerated. The lesions bleed easily especially when the surface crust is removed. This particular carcinoma most commonly appears on the face and forehead but is not unknown on the arms, legs or trunk. It is caused by excessive exposure of the skin to sun and is the most common form of skin cancer.

Squamous cell carcinoma

This is entirely different from the basal cell carcinoma. It is most frequently seen in elderly people. The onset is gradual. Clinical signs are the presence of a warty surface with ulcerations occurring when the growth is 1–2 cm in diameter. The edge of the ulcer thickens. It is a true invasive tumour that develops in normal tissue or in a pre-existing lesion.

Light-induced skin disease

There are a number of skin diseases that develop as a result of exposure to sunlight without the use of a protective sunscreen. Damage to the skin may take some time to become obvious.

Some of the problems that arise are non-cancerous, such as premature ageing due to damage of the collagen and elastin content of the skin, increased pigmentation as a result of exposure to sunlight or use of sun beds when wearing perfume, sunburn and reactions to certain plants and photosensitive drugs such as Prozac. The late 1990s highlighted an increase of sun induced skin cancers. In the summer of the year 2000 statistics showed that over 40 000 cases of skin cancer were reported, with 2000 of these being terminal.

Conclusion

There are very many skin diseases and disorders that professional electrolysists may come into contact with during the course of their work. It will be possible to treat some of these conditions with electrical epilation, whereas others may be contraindicated, and a number may indicate referral to the medical profession. Many clients expect their electrolysist to be a walking encyclopaedia when it comes to skin lesions and disorders. It is essential that every electrolysist becomes familiar with the most common conditions and is able to decide which course of action will give most benefit to the client. As far as skin disorders and lesions are concerned, 'when in doubt, don't treat!'

Review questions

1 Name the internal and external influences that affect the health and condition of the skin.
2 Define the term 'skin lesion'.
3 Describe the following:
   (a) Macule
   (b) Pustule
   (c) Plaque
   (d) Vesicle
   (e) Fissure.
4 Explain the meaning of the term 'hyperkeratosis'.
5 What is the difference between hyperpigmentation and hypopigmentation?
6 Name six causes of skin disease.
7 Briefly describe the appearance of:
   (a) Impetigo
   (b) Herpes zoster
   (c) Pityriasis versicolor.
8 Describe the differences between:
   (a) Sebaceous cyst
   (b) Milia
   (c) Xanthoma.
9 Define the terms 'eczema' and 'dermatitis'.
10 Describe the appearance of allergic dermatitis.
11 What is the difference between allergic dermatitis and contact dermatitis?
12 Describe the clinical signs of psoriasis and name the aggravating factors.
13 Compare acne vulgaris with rosacea.
14 Give the characteristics of a pigmented naevus.
15 Give the clinical characteristics of a rodent ulcer.

3 Hair

Hair has made its presence felt for thousands of years. For many people it is their crowning glory, yet for so many others it is a source of great distress and embarrassment. In the opening years of the twenty-first century, hair removal has grown into a multi-million pound business. Both men and women are looking for a permanent method of hair removal and new technology is developing rapidly. Yet electrolysis, established since 1875 and now entering into a third century, has shown over the years that permanent hair removal can be achieved.

To eliminate hair permanently, it is essential the electrolysist is familiar with both the structure and the growth cycle of hair. This chapter explains both in some detail, and a Glossary is included at the end of the chapter to help with unfamiliar terms.

Structure of the hair

Hairs are keratinized structures growing out of hair follicles, which are sac-like indentations of the epidermis. Keratin is a hard, horny substance, which resists digestion by pepsin and is insoluble in water, organic substances, dilute acids and alkalis. Keratin is composed of a combination of hydrogen, oxygen, sulphur and nitrogen.

The terminal hair can be divided into three separate parts:

- The shaft – this is the portion above the skin's surface.
- The root – the portion lying in the follicle.
- The bulb – the enlarged base of the root.

Types of hair

There are three distinct types of hair:

- Lanugo – primary.
- Vellus – secondary.
- Terminal – tertiary.

Lanugo hair is usually found in foetal life and is normally shed around the seventh to eighth month of pregnancy, to be replaced by vellus or terminal hair. This type of hair is long, fine, downy and soft in texture, without a medulla and usually does not contain pigment.

Vellus hair is short, fine, downy, and soft and does not contain pigment. It is found on the body generally and rarely exceeds 2 cm in length. These hairs do not contain a medulla. The base of the vellus hair lies very close to the skin's surface. Vellus hairs do not become terminal hair unless stimulated by topical or systemic conditions.

Terminal hairs are deeper, longer and coarser than vellus hairs. They contain pigment and vary in shape, diameter, texture, length and colour. These hairs are found at specific sites in the body and can be divided into three groups:

1. *Asexual* – genetic hair present at birth. Asexual hair refers to hair found on the scalp, eyebrows, and eyelashes and to a lesser extent on the

forearms and legs in both sexes of all ages. Although this hair type is influenced by growth hormone production, steroids do not influence growth.

2 *Ambisexual* – develops in both sexes at puberty. This type of hair growth is influenced by the increased gonadal and androgen production. Areas where ambisexual hair growth occurs are the axilla, pubis, lower limbs and abdomen in both sexes. Growth on the forearms and legs becomes more profuse at this stage of life. The density and rate of growth differs widely between sexes, individuals of the same sex and various body sites.
3 *Sexual hair* includes the beard, moustache, nasal passages, ears and external body hair, e.g. back and chest. Sexual hair is influenced by increased androgen hormone production by the gonads. Testosterone levels in men are 15–30% higher than in women. Sexual hair is more pronounced in men due to the higher levels of progesterone production by the testes.

The shape of the follicle determines the shape of the hair, i.e. straight hairs grow from straight follicles whereas curly or wavy hairs grow from curved follicles. Hair, which is kinked or frizzy grows from follicles that have become distorted at the base as a result of mechanical interference such as waxing or tweezing.

Terminal hair is composed of three layers (see Figure 3.1).

1 *The cuticle* is composed of a single layer of scale-like cells, which point towards the tip of the hair. These cells overlap like the tiles on a roof so preventing the passage of foreign objects into the follicle. This overlapping allows the cuticle of the follicle to interlock with the cuticle

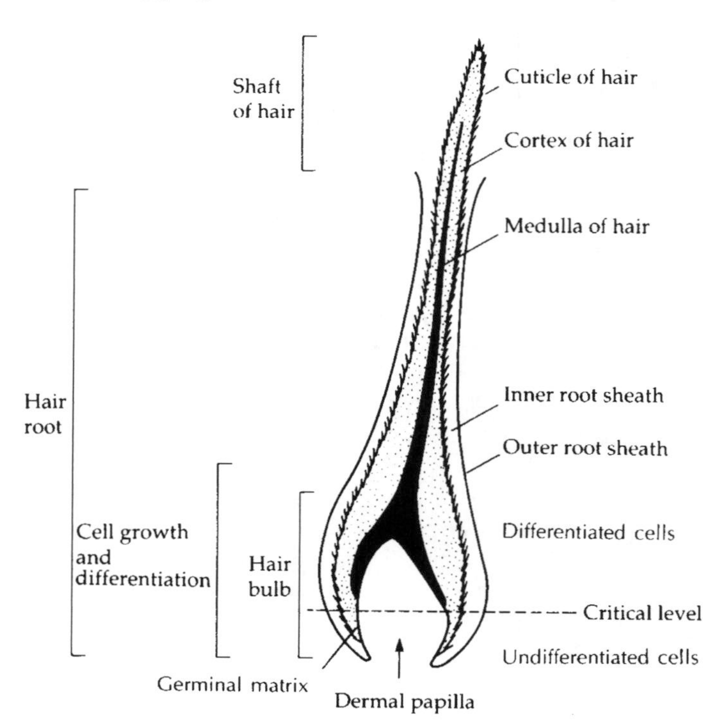

Figure 3.1
Longitudinal diagram of terminal hair

of the hair, so holding the hair in place. A thin layer of lipids (fatty substances) and carbohydrates surrounds the cuticle and may protect the hair from the effects of physical and chemical agents. This layer does not contain pigment. The function of the cuticle is to confine and protect the cortex and give the hair its elasticity.

2 *The cortex* lies inside the cuticle and forms the bulk of the hair. It consists of elongated keratinized cells cemented together. Melanin granules contained within this layer give the hair colour. A number of air spaces are contained within the cortex. In the living part of the hair these spaces are filled with fluid, which gradually dries out as the hair grows. These spaces are larger at the base of the hair, becoming smaller at the tip.

3 *The medulla,* when present, is found in the centre of the hair. It may be continuous or discontinuous and may vary within the same hair. The medulla is formed of loosely connected, keratinized cells. Air spaces in the medulla determine the sheen and colour tones due to reflection of light.

The hair follicle

The hair follicle is a downward extension of the epidermis of the skin. Hair follicles, together with the sebaceous gland, form the pilo-sebaceous system. Attached to the follicle below the sebaceous gland is the arrector pili muscle. It is the contraction of this muscle that causes the hair to stand on end – so giving the goose-pimpled effect.

The hair follicle consists of the following structures (see Figure 3. 2):

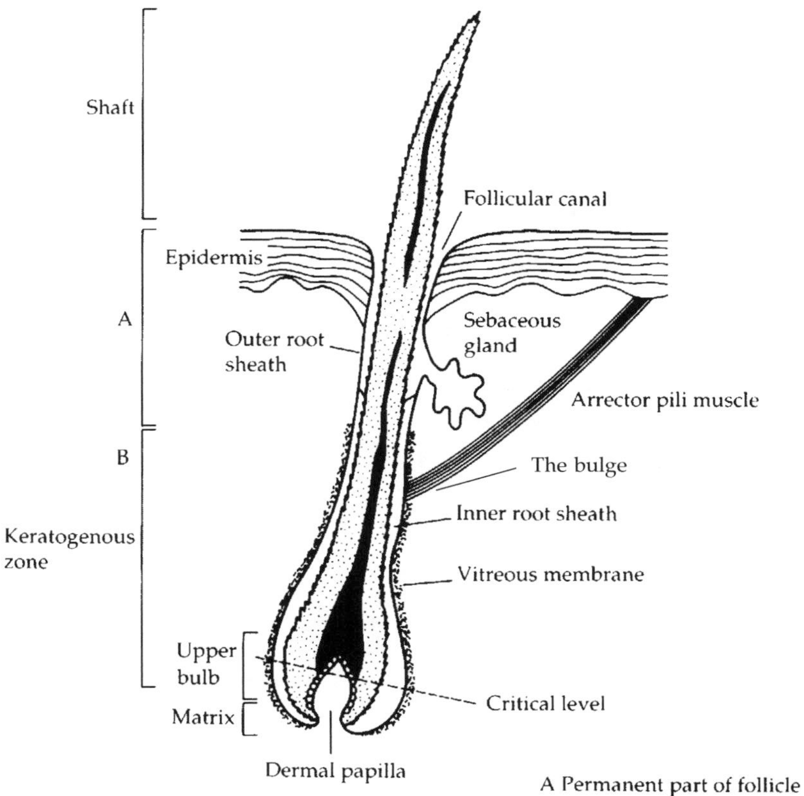

Figure 3.2 Anagen hair in its follicle

- Inner root sheath.
- Outer root sheath.
- Vitreous membrane.
- Connective tissue.
- The bulge.

The inner root sheath holds the hair in the follicle by interlocking with the cuticle of the hair to the level of the sebaceous gland. The inner root is composed of three distinct layers:

1. The innermost layer is the cuticle, which interlocks with the cuticle of the hair.
2. Huxley's layer is the middle layer and is the thickest of the three layers.
3. Henle's layer is the outer layer and consists of a single layer of cells.

The inner root sheath (see Figure 3.3) originates from the base of the follicle, growing up in unison with the hair until its reaches the level of the sebaceous gland. The hair then continues to grow up, on its own, through the (follicular) hair canal.

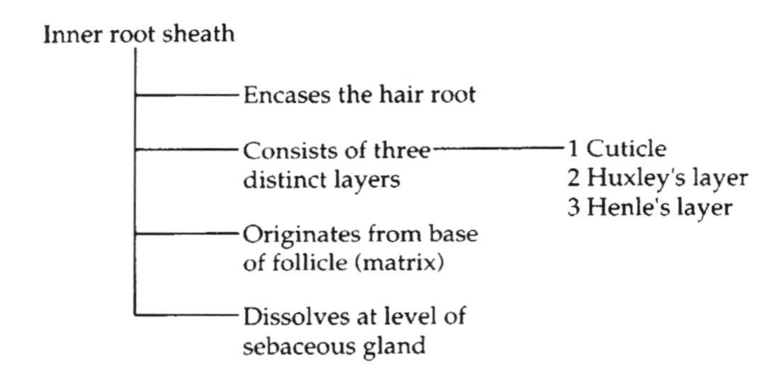

Figure 3.3 The inner root sheath

The outer root sheath surrounds the inner root sheath, and is continuous with the mitotic layer of the epidermis. At the level of the sebaceous gland, the cellular structure of this layer cannot be distinguished from that of the surface epidermis. Large amounts of water and glycogen are found in this layer, the highest concentration being found in the cells between the neck of the bulb up to the level of the sebaceous glands. The thickness of this layer is uneven and, unlike the inner root sheath, it does not grow up in unison with the hair. The outer root sheath is the permanent source of the *hair germ cells* from which new follicles develop when stimulated by circulating hormones and enzymes. It is, therefore, important that these cells are destroyed during treatment to prevent further growth.

The hair bulb

The hair bulb can be divided into upper and lower regions. The lower region contains the undifferentiated cells, whereas the cells in the upper region differentiate to form the hair and inner root sheath.

An imaginary line drawn across the widest part of the hair bulb would separate the two regions and is known as the critical level. Below the

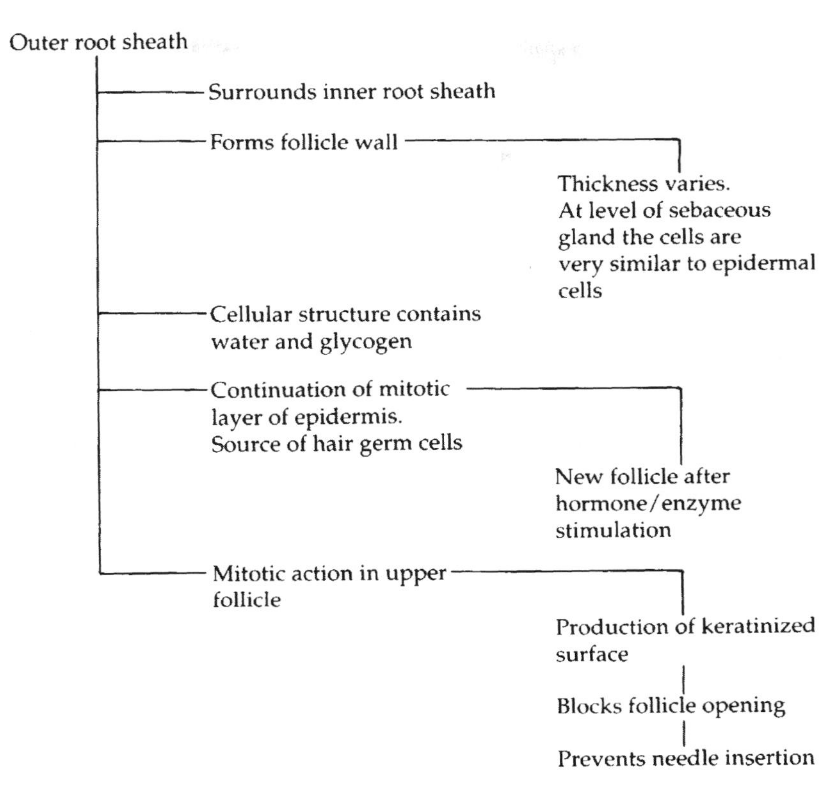

Figure 3.4 The outer root sheath

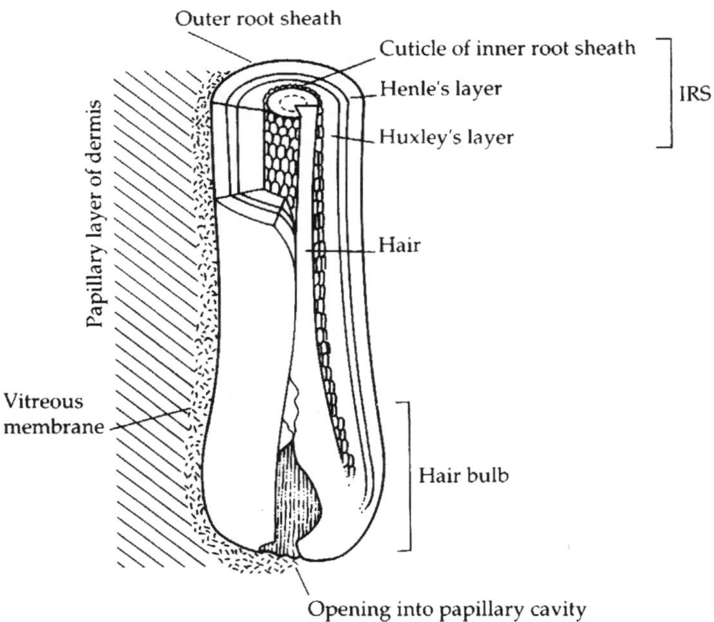

Figure 3.5 The bulb of the follicle

critical level is the germinative centre or matrix, where all the cells are mitotically active. From the matrix, cells move up to the upper part of the bulb where they elongate vertically and increase in volume.

The keratogenous zone is found in the topmost part of the bulb, above the critical level, terminating approximately one-third of the way between the tip of the papilla and the skin's surface. Melanocytes are contained within the upper part of the bulb and are concerned with pigmentation of the hair. The matrix contains very little melanin, therefore showing the separation of the upper from the lower bulb very clearly.

The upper bulb can be compared to the spinous layer of the epidermis. In both places indifferent epidermal cells become larger, acquire pigment, synthesize fibrous proteins, become reoriented and undergo the final stages of keratinization. About halfway up the follicle is the keratogenous zone where keratinization takes place.

The connective tissue sheath

The connective tissue surrounds the outside of the sebaceous gland. It is a continuous extension of the papillary layer of the dermis and includes the dermal papilla. The dermal papilla is the main source of sustenance for the entire follicle structure. The vitreous membrane, which is a hyaline, non-cellular glossy membrane, lies next to the outer root sheath. Surrounding the vitreous membrane are two layers of connective tissue. The first layer consists of compact fibres arranged circularly around the follicle. The second layer is composed of longitudinal bundles of connective tissue, attached at the base of the dermal papilla by a stalk.

The bulge is the area to which the arrector pili muscle is attached. The 'bulge' region is believed to be the storage area for hair follicle stem cells. Hair follicles go through a cycle of growth and rest. With each renewed attempt to produce hair, the hair follicle must obtain a source of cells to form the matrix cell population that make hair. The source of these cells is believed by some dermatologists to be the bulge region. Other dermatologists suggest that stem cells are not present in the bulge region at all and that new matrix cells are obtained from the root sheath (Y. Naraisawa and Hiromu Kohda (1996) Extracts from *Archives of Dermatological Research*, Volume 288, Issue 2).

The nerve supply to the follicle

The nerve supply to the skin is contained within the dermis and around the hair follicles, reaching the base of the epidermis. It is this dermal network of nerves to the hair and follicle which forms the skin's sensory nerve supply. The nerve plexus surrounds the hair follicle extending to the upper part of the pilary canal with a few fibres reaching the sebaceous glands. During *telogen*, the follicle is shorter, therefore the nerve supply will be closer to its base.

Tactile stimuli increase the range of sensitivity to minute mechanical disturbances, due to the sensory nerves of the hair follicle.

The blood supply to the follicle

A network of capillaries, the density of which differs between active and resting follicles, surrounds hair follicles. This network of capillaries surrounds the entire follicle, together with the sebaceous gland. A loosely woven network of capillaries is formed above the level of the sebaceous gland. This network extends to, and is continuous with, the loops of capillaries found in the *papillary plexus*. These loops of capillaries form a vascular ring around the terminal part of the pilary canal. In the *dermal papilla*, capillaries from a central tuft of vessels extend to the walls of the inner surface of the follicle and practically come into contact with it. The vascular system of each follicle, including the plexus around the sebaceous glands, is a continuous unit (see Montagna and Ellis, *Structure and Function of the Skin*).

The vascular system becomes smaller in proportion to the size of the follicle. The lower part of the follicle containing vellus hair has very few capillaries, with no vessels penetrating the dermal papilla.

A number of changes take place during *catagen*. While the vascular system around the bulb and dermal papilla remain intact during early catagen, the vitreous membrane and connective tissue sheath become thicker and wrinkled. When the outer root sheath and bulb collapse during late catagen, the blood vessels of the lower plexus remain intact. When the follicle shortens, the papillary vessels retreat upwards and the vessels of the papilla lose their clear outline.

Some capillary tufts in the dermal papilla collapse and show the first sign of degenerative changes in the vascular system of the follicle. In advanced catagen the lower follicle shrivels, retreating upward, leaving a trail of connective tissue behind, known as the *epidermal cord*. The dermal papilla, freed from the bulb, remains in contact with the retreating follicle via the epidermal cord.

All these changes take place in the lower vascular network of the follicle, which remains relatively intact even in advanced changes. Some of the capillaries of the lower network degenerate when the lower third of the follicle is reduced to a thin, long strand of cells. At the completion of catagen, most of the follicle below the bulge is reduced to a hair germ, at the base of which the dermal papilla is attached.

Around the sebaceous gland and funnel-shaped entrance (infundibulum) of the pilary canal, the upper follicle network remains intact and is similar to that around the active follicles. When the resting follicle becomes active again, the new bulb must advance through the collapsed bundle below the dermal papilla, growing inside it. The major vessels of the follicle remain intact during catagenic changes.

Exchange of nutrients takes place through the wall of the bulb that faces the dermal papilla. Not all dermal papillae are equally supplied with vessels, and the amount of vascular tissue in a papilla is related to the size of the follicle. The wider the diameter of the follicle, the larger the capillary tuft it contains. The dermal papillae of large, active human follicles are very wide and contain a large number of capillaries.

The hair growth cycle

The growth cycle of the hair follicle can be compared to that of a flower bulb, for example, the daffodil or tulip (see Figure 3.6). During winter the flower bulb is dormant, without roots or leaves. In spring, the roots begin to form, growing down into the soil, the leaves and flower bud begin to form within the bulb; and there is a gradual move up, towards the surface of the soil. The leaves push up through the ground, shortly followed by the flower. When the flower has opened fully, it begins to die, and the whole cycle begins again. So it is with the follicle.

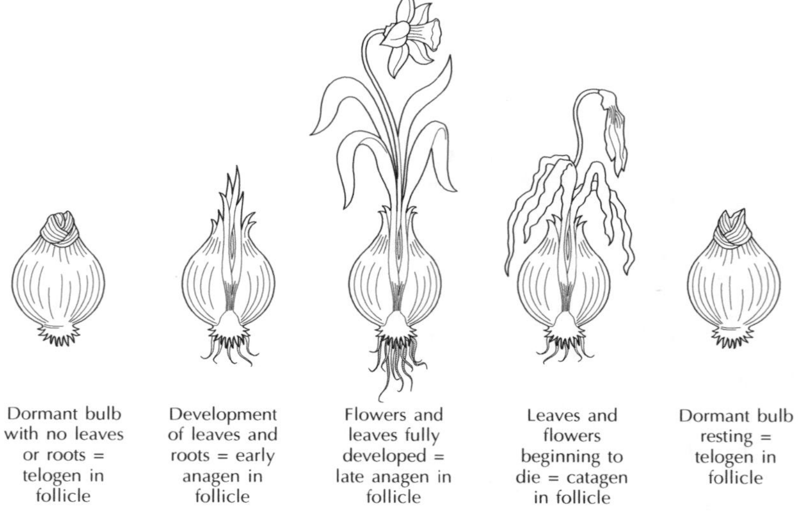

Figure 3.6 Comparison of the hair growth cycle with that of a flower (daffodil) bulb

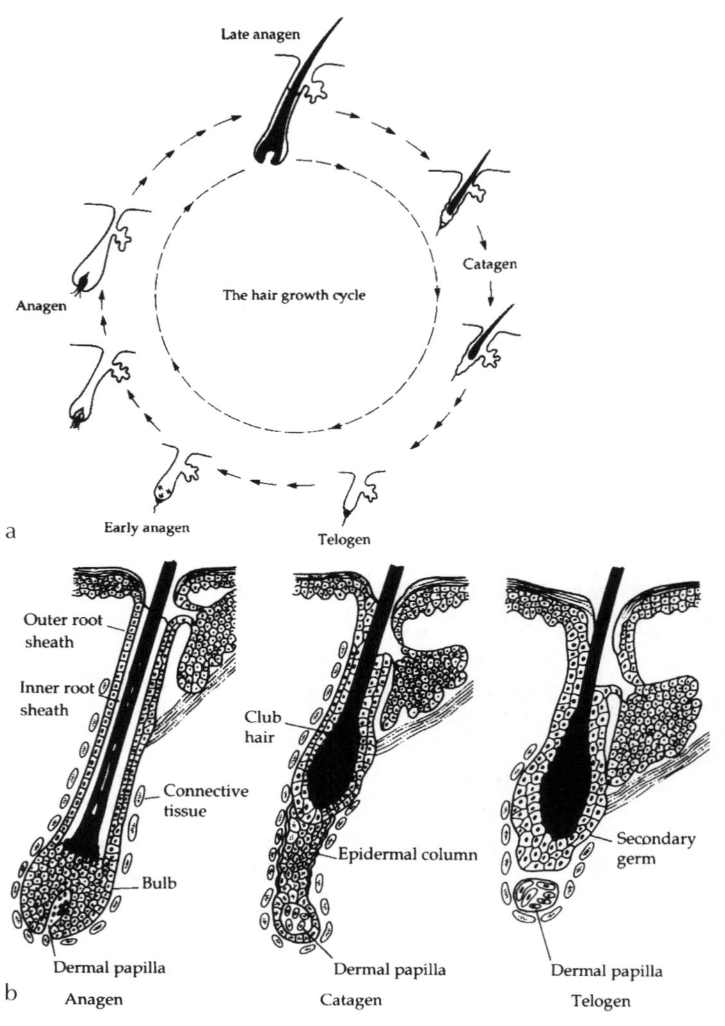

Figure 3.7 The hair growth cycle

The follicle goes through three distinct stages of development (see Figure 3.7). These stages are known as:

1 Anagen – active.
2 Catagen – transitional.
3 Telogen – inactive or dormant.

Detailed knowledge of the hair growth cycle can be attributed to a number of people, in particular Dry (1926), Ligman (1959) and Montagna and Parakkal (1974). Some current reference works on the subject can be found in the Bibliography.

Anagen

Anagen is the active stage, which results in the complete restructuring of the lower follicle. Hair germ cells contained within the dermal cord begin to multiply by mitosis. The dermal cord grows downwards into the dermis, at the same time growing in width, until the bulb, which has formed at the tip of the dermal cord, engulfs the dermal papilla.

The specialized nature of papilla cells is maintained even when papillae are isolated from the epidermal cells. It has been shown (Orfanos, Montagna and Stuttgen, 1981) that when direct papilla/epidermal contact is maintained the papilla induces the following changes:

- The development of a matrix.
- Increased mitotic activity of the epidermis.
- Hair growth.
- Follicle lengthening.

Before the follicle has reached its full depth, the mitotic cells of the *germinal* matrix, found in the lower part of the bulb, become active. These cells move upwards, differentiating into cells that produce (a) the hair and (b) the inner root sheath.

Towards the end of the anagen stage thinning and lightening of the pigment at the base of the hair shaft takes place, and melanin production stops. Melanocytes reabsorb their dendrites. During anagen the follicle receives its nourishment through the dermal papilla.

Oliver (1966) and others (e.g. Jahoda, 1992) revealed that the removal of the dermal papilla stops hair growth but that the lower third of the dermal sheath is capable of supplying new cells for regeneration of a new dermal papilla, by infiltrating and transforming at the site of the original dermal papilla with subsequent hair follicle regrowth. When the lower third of the hair follicle is removed, restructuring of the dermal papilla is unable to occur and the follicle is permanently destroyed.

Catagen

This stage follows on from anagen. It is known as the transitional stage where the club hair is formed. The dermal papilla *separates* from the matrix and the hair starts to rise in the follicle. At this time the hair is held in the follicle by the cells of the inner root sheath and receives its nourishment from the follicle wall. As the follicle begins to shrink and collapse below the rising hair, the epidermal cord is formed from undifferentiated cells. The hair becomes drier, losing water and glycogen.

Telogen

Telogen is the final phase in the hair growth cycle. It is known as the resting or dormant stage. The follicle remains inactive or dormant until stimulated into anagen stage, when the whole process is repeated.

Telogen only lasts for a few weeks, with the club hair often being retained within the follicle until the new hair is produced. The club hair is then pushed out of the follicle as the new hair grows. The telogen follicle is one-third to half the length of a full anagen follicle.

Terminal hairs vary in texture, pigment content, colour and density from one body area to another and between different individuals and nationalities.

Summary

To summarize, the hair is a keratinized structure consisting of three layers – cuticle, cortex and medulla – which grows from the hair follicle. The follicle forms part of the pilo-sebaceous unit, which reaches from the epidermis down to the dermal papilla situated in the dermis. The hair follicle goes through a growth cycle divided into three phases, these being anagen–active, catagen–transitional and telogen–resting/dormant. The duration of these phases is summarized in Table 3.1.

Table 3.1 Hair growth table

Body area	*% Resting hairs (telogen)*	*% Growing hairs (anagen)*	*Duration of telogen*	*Duration of anagen*	*No. of follicles per cm²*	*Daily growth rate*	*Approx. depth of terminal anagen follicle*
Scalp	13	85	3–4 mth	2–6 yr	350	0.35 mm	3–5 mm
Eyebrows	90	10	3 mth	4–8 wk		0.16 mm	2–2.5cm
Ears	85	15	3 mth	4–8 wk			
Cheeks	30–50	50–70			880	0.32 mm	2–4 mm
Beard (chin)	30	70	10 wk	1 yr	500	0.38 mm	2–4 mm
Upper lip	35	65	6 wk	16 wk	500		1–2.5 mm
Axillae	70	30	3 mth	4 mth	65	0.30 mm	3.5–4.5 mm
Trunk					70	0.30 mm	2–4.5 mm
Pubic area	70	30	12 wk	4 wk	70		3.5–4.75 mm
Arms	80	20	18 wk	13 wk	80	0.30 mm	
Legs and thighs	80	20	24 wk	16 wk	60	0.21 mm	2.5–4 mm
Breasts	70	80			65	0.35 mm	3–4.5 mm

Source: Cosmetic and Medical Electrolysis and Temporary Hair Removal (R.N. Richards and G.E. Meharg, Medric Ltd, 1991)

The follicle structure consists of inner and outer root sheaths, the outer root sheath being separated from the connective tissue by a vitreous membrane. Both the follicle and hair receive nourishment from the dermal papilla. Circulating hormones and enzymes within the blood stimulate development and growth.

Glossary

Hair Keratinized structure that grows out of the hair follicle.
Anagen Active stage of growth where lower follicle rebuilds and new hair is formed.

Asexual hair Growth not governed by hormones, e.g. scalp, eyebrows, eyelashes.
Catagen Follows anagen. Hair separates from the dermal papilla. Club hair is formed. Lower follicle begins to shrivel and collapse.
Club hair Develops in catagen. The bulb of the hair dries out and becomes brush-like. The club hair is held in the follicle by the cells of the inner root sheath.
Epidermal cord Slender cord of hair germ cells which enables the retreating follicle to maintain contact with the dermal papilla.
Hair follicle Sac-like indentation of the epidermis that grows down to the subjacent dermis.
Hair germ Consists of undifferentiated cells, which produce new hair when stimulated by circulating hormones and enzyme action.
Infundibulum Funnel-shaped opening to the follicle.
Keratin Hard, horny substance made up of carbon, hydrogen, sulphur, oxygen and nitrogen.
Keratogenous zone Area where keratinization takes place in the hair follicle. This is found in the upper part of the bulb and finishes approximately one-third of the way between the tip of the papilla and the skin's surface.
Pilary canal Consists of upper third portion of the outer root sheath, which extends above the entrance of the sebaceous gland.
Pilo-sebaceous unit Formed from the hair follicle and the sebaceous gland.
Sexual hair Growth and development influenced by hormones, particularly androgens. Found in the axilla and pubis.
Telogen Final stage of hair growth cycle. Follows catagen. Follicle is inactive or resting and a half to one-third the length of the anagen follicle.

Review questions

1. Name the three separate parts of a terminal hair.
2. Names the three types of hair.
3. Give a detailed description of a terminal hair.
4. Describe the structure of the hair follicle in the anagen stage.
5. What is a hair follicle?
6. Name the structures of an anagen follicle.
7. What is the dermal cord?
8. How does the dermal papilla maintain contact with the follicle?
9. Describe the following: outer root sheath, inner root sheath, hair bulb, connective tissue, and vitreous membrane.
10. Name the three stages of the hair growth cycle.
11. Describe the three stages of follicle development and hair growth.
12. What is the function of the arrector pili muscle?
13. What does the 'critical level' mean?
14. Where is pigment found in the hair?
15. Explain the difference between a terminal and a vellus hair.
16. How does the hair receive its nourishment?
17. Define 'keratogenous zone'.
18. What is the function of the germinal matrix?

4 The cardiovascular system

It is not the intention here to cover the cardiovascular system in any depth – there are many good anatomy and physiology books that deal with the subject well. The main aim is to give a general insight into the functions and clotting mechanism of blood, so that the electrolysist will understand the role of the cardiovascular system in relation to the following:

- Healthy skin and hair.
- The endocrine system.
- The clotting mechanism at the site of an injury arising out of accidental misprobe or during the treatment of naevi and capillaries.

The human body is composed of a number of organized systems, none of which can function independently. The cardiovascular system acts as the transport medium, carrying oxygen, nutrients and hormones etc. to the tissue cells while at the same time removing carbon dioxide and waste products via the blood.

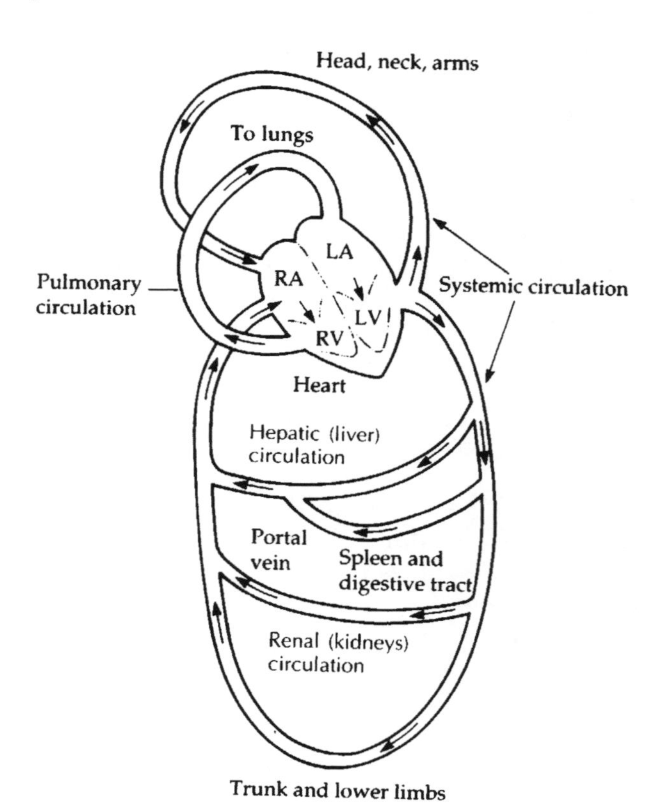

Figure 4.1 General circulation

Blood is carried through the system via a series of arteries, arterioles, capillaries, venules and veins, with the heart being the receiving and distribution centre (see Figure 4.1).

The circulatory system can be divided into the:

1 The *blood circulatory system*, consisting of:
   (a) the heart
   (b) the blood vessels:
      arteries
      arterioles
      capillaries
      venules
      veins.
2 The *lymphatic system*, consisting of:
   (a) lymph nodes
   (b) lymph vessels.

The efficient transport and function of blood within the body is essential in order to maintain health and well-being. Blood is transported around the body by means of the heart, which acts as a pump, the arteries and veins.

The heart

The heart is a strong, muscular pump formed from three layers of muscular tissue. These are (i) the pericardium, (ii) the myocardium and (iii) the endocardium.

The *pericardium* is separated into two sacs, the outer fibrous sac and the inner serous membrane. Serous fluid is secreted into the space between the inner and outer layers, so allowing smooth movement between the two.

The myocardium is lined by the *endocardium*, a thin, smooth membrane. The *myocardium* is specialized cardiac muscle which functions continuously and is not under control of the will. The muscle is thickest at the apex and thinner at the base.

The heart is divided into four chambers – two upper and two lower. The septum separates the left side of the heart from the right (see Figure 4.2).

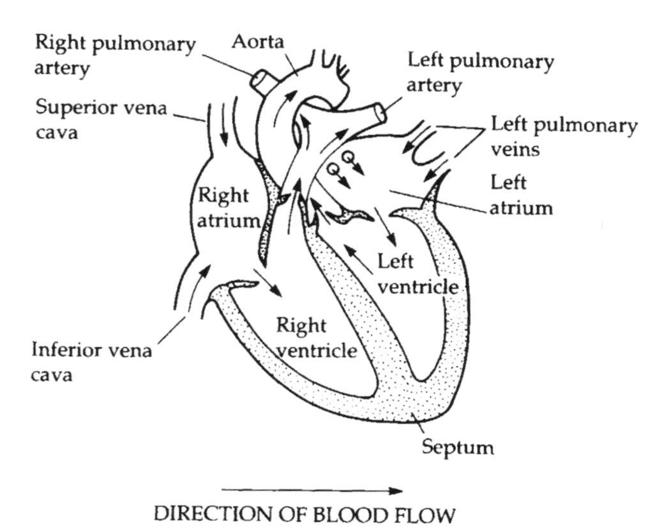

Figure 4.2 The heart and associated blood vessels

The four chambers are known as the right and left ventricle and the right and left atrium. The heart receives dark red, deoxygenated blood into the right atrium from the superior and inferior venae cavae. This blood then passes through the right atrioventricular (tricuspid) valve into the right ventricle. From here the blood is carried via the two pulmonary arteries to the lungs where the exchange of oxygen and carbon dioxide takes place. Blood is returned from the lungs to the left atrium by the two pulmonary veins. After passing through the left atrioventricular (tricuspid) valve into the left atrium, the blood leaves the heart through the aorta for distribution around the body via the arteries.

Arteries

Arteries are strong muscular tubes that carry oxygenated blood, hormones, nutrients and other substances to the capillary network for distribution to the tissue cells. The pulmonary artery is the exception to this rule.

The function of the arteries is to pump oxygenated blood around the body, with the exception of the pulmonary artery, which takes deoxygenated blood from the heart to the lungs. Blood in the arteries moves in spurts owing to the pumping action of the heart. Contractile muscular walls also help circulation of blood between pulses.

Arterioles

Arterioles (or pre-capillaries) connect arteries to capillaries. The structure is similar to that of arteries but the walls are much thinner.

Capillaries

The arterioles gradually reduce in size to form capillaries, minute in size, which can be found in nearly every tissue of the body. Their diameter differs in different parts of the body. Capillaries connect arterioles to venules. The density of capillary networks varies in different areas, being denser in areas such as the dermis. Exchange of gases, nutrients, hormones and metabolic waste takes place in the capillary network.

Venules

Venules are larger in structure than capillaries and form the link between capillaries and veins.

Veins

Veins are similar in structure to arteries except the walls are thinner, less muscular and they contain valves that prevent the backward flow of blood.

Composition of blood

Blood is a salty-tasting fluid consisting of plasma and solids. Plasma is an alkaline, straw-coloured substance composed of 91% water, 8% protein and 0.9% salts. The solids consist of:

1 Platelets or thrombocytes, which play a part in the control of bleeding after an injury and the clotting procedure.
2 Erythrocytes or red cells carrying haemoglobin, which combine with oxygen to form oxyhaemoglobin;
3 Leucocytes or white cells whose function is to ingest bacteria and protect the body against micro-organisms, thereby fighting infection.

Function of blood

The function of blood is to act as a transport medium for the following:

- nutrients, tissue salts and enzymes } to tissue cells
- oxygen } to tissue cells
- hormones – from the endocrine glands to the target organs
- urea } from the cells to the elimination organs
- uric acid } from the cells to the elimination organs
- carbon dioxide } from the cells to the elimination organs
- antibodies – to fight infection
- drugs and medication.

Blood is also concerned with the body's temperature control by means of vasoconstriction and vasodilatation of the surface capillaries. When too much heat is present, the capillaries dilate, so releasing heat to the skin's surface. Conversely, when the body is cold, capillaries constrict, so taking blood away from the skin's surface thereby containing heat within the body.

Platelets, together with other substances within the blood are responsible for removing bacteria and micro-organisms from the blood, so helping to fight infection. When a foreign body such as an epilation needle or micro-lance pierces the skin, the tip penetrates the surface, allowing germs and bacteria to enter the bloodstream. Blood brings white cells to the area, their function being to kill the bacteria and germs. The site of the injury becomes red, swollen, painful and hot owing to the increased local blood supply and the activity of the cells in the area. This particular injury is known as 'needle stick injury'. It is possible to transmit hepatitis B and HIV in this manner if contaminated needles are used.

The blood clotting mechanism

When the skin is cut or injured, blood appears at the surface. After a short time, a clot should form over the area to seal the skin, so producing a scab. The scab will eventually fall off when the skin underneath has healed. The exception to this is the condition of haemophilia, where clotting does not take place. A severe case could lead to death of the person concerned.

The clotting process is dependent upon a number of factors (see Figure 4.3).

The health and efficiency of the blood vascular system can be affected by:

1. Badly balanced or incorrect nutrition.
2. Smoking.
3. Anti-coagulant drugs such as warfarin and aspirin.
4. Alcohol.

When the blood is not able to do its job efficiently the healing rate of the skin after electro-epilation will not be good. In clients who smoke regularly, or who perhaps take an aspirin immediately before the treatment of telangiectasia or spider naevi, the blood will not coagulate well. Clients who follow a badly balanced diet that lacks sufficient vitamins and minerals, or who eat irregularly, may find that the skin is not as healthy and may take longer to heal after treatment.

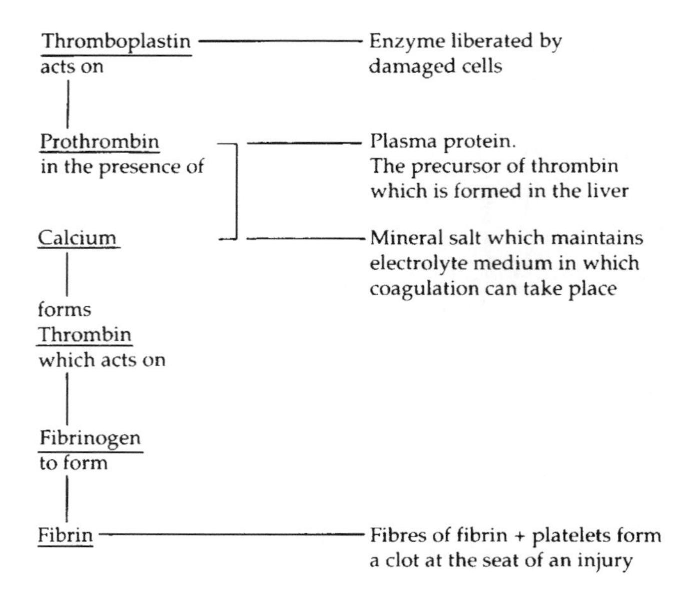

Figure 4.3 Factors necessary to the clotting process

Review questions

1. Draw and label a diagram of the heart.
2. Describe the differences between arteries and veins.
3. What is the composition of blood?
4. What is meant by the blood-clotting mechanism?
5. Name four factors that affect the health and efficiency of the blood vascular system.

5 The endocrine system

The endocrine system can be referred to as the 'orchestra' of the body, with the conductor being the pituitary gland in conjunction with the hypothalamus, which forms a link between the endocrine and nervous systems. When this system is working in harmony, the balance of hormone levels will be correct. This in turn will enable all the other systems of the body to carry out their functions effectively. Body growth, development and functioning are all dependent on the efficiency of the endocrine system. Too much or too little of any hormone will cause an imbalance resulting in certain disorders or malfunction.

The process involved in the endocrine system is shown in Figure 5.1.

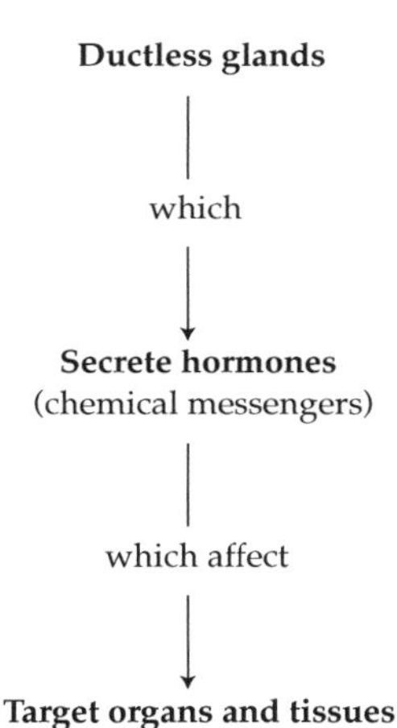

Figure 5.1 The action of the endocrine system

Hormones

Hormones are complex chemical substances produced by endocrine glands. Their function is to stimulate or inhibit the action of specific glands, organs or tissues. Hormones are released directly into the bloodstream. They are slow-acting chemical messengers which control functions such as metabolism, growth of body tissues and cells, also mental and physical coordination.

The hypothalamus, pituitary and target glands control the hormone levels in the blood by means of the feedback mechanism. When the level of a particular hormone falls, the hypothalamus notifies the pituitary gland to increase secretion of the trophic hormone to stimulate hormone production by the target gland, e.g. low levels of thyroxine trigger the hypothalamus into informing the pituitary of the need to increase its output of the thyroid stimulating hormone (TSH), which will then influence the thyroid gland to increase secretion of thyroxine. In reverse, when there is a high level of a particular hormone, the hypothalamus will inform the pituitary to decrease secretion of a trophic hormone (see Figure 5.2).

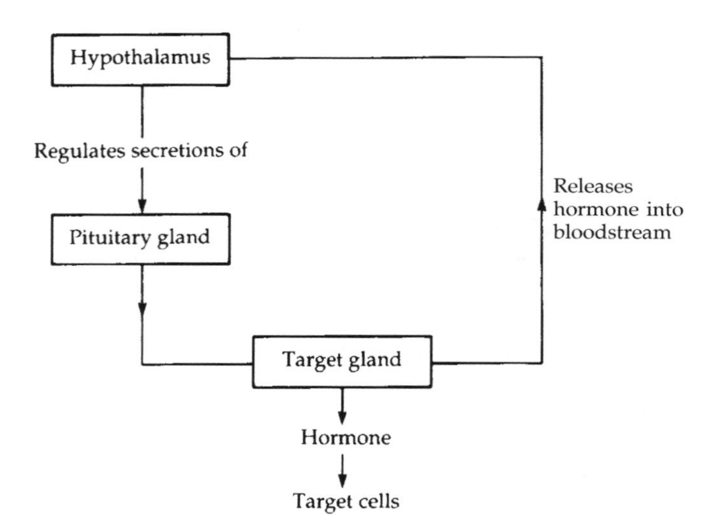

Figure 5.2 The feedback mechanism

Without the correct balance of hormones it is not possible to lead a healthy life. Stress control, body growth, sexual development, temperature regulation and memory are some of the functions that are dependent on the presence of hormones.

Endocrine glands

The endocrine glands are situated in specific sites of the body. They are ductless glands which produce *hormones,* secreting them directly into the bloodstream. These glands of internal secretion work closely with the nervous system to form the communication network of the body (see Figure 5.3).

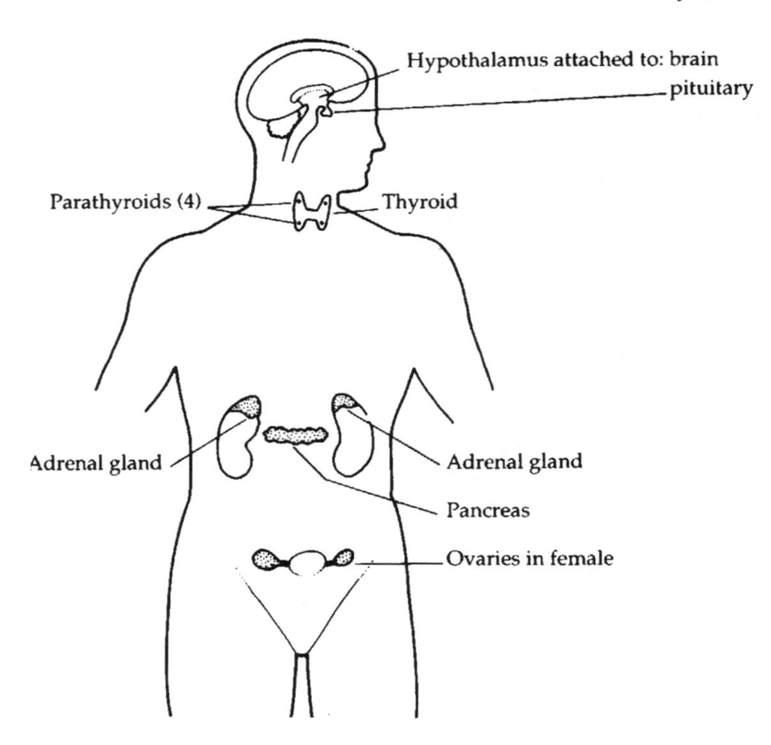

Figure 5.3 Positions of the endocrine glands

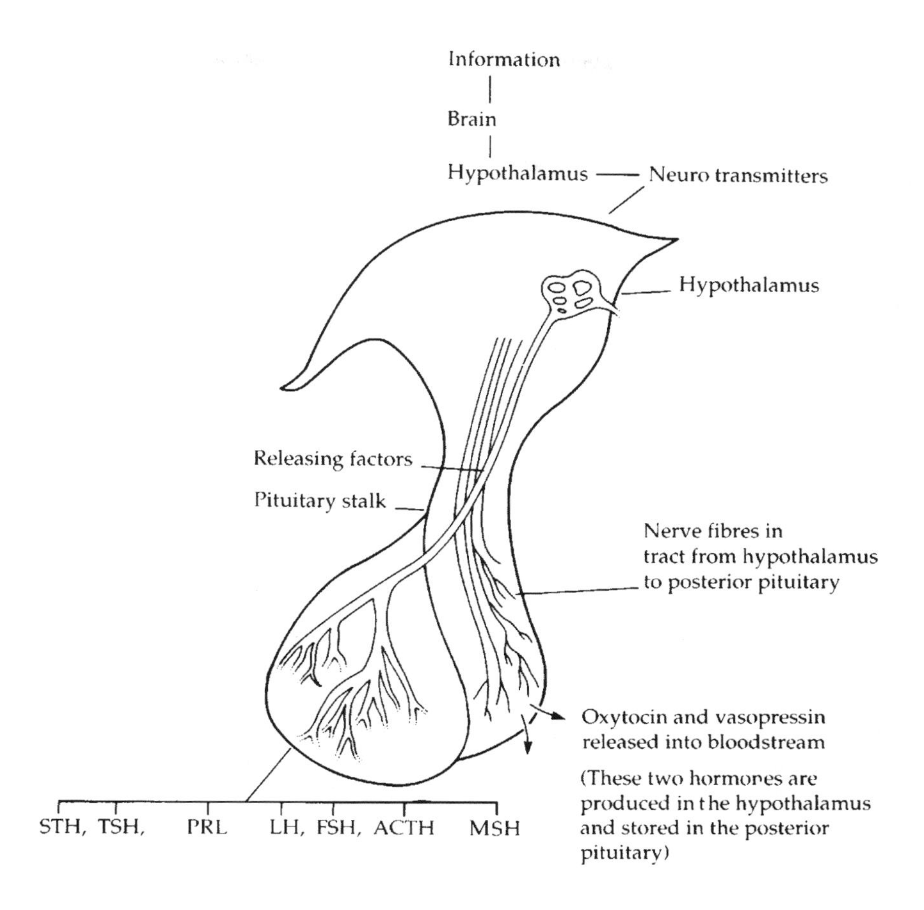

Figure 5.4 The hypothalamus

The glands differ in shape, size and location, although their function is the same: to produce hormones.

The hypothalamus

The hypothalamus is part of the mid-brain and is situated between the thalamus and the pituitary gland. It links the cerebral cortex of the brain with the pituitary by means of a stalk richly supplied with nerve fibres and blood vessels.

The hypothalamus produces releasing and inhibiting hormones which directly affect the pituitary gland. The releasing hormones stimulate the release of trophic hormones into the bloodstream, whereas inhibiting hormones prevent the release of prolactin and melanocyte-stimulating hormones. Vasopressin and oxytocin are both produced by the hypothalamus and stored in the posterior pituitary gland.

The role of the hypothalamus is coordination of the endocrine and autonomic nervous systems. It is responsible for the control of:

1 The autonomic nervous system.
2 Metabolic processes.
3 Secretion of pituitary hormones.
4 Sleep.
5 Appetite.

6 Regulation of sexual function.
7 Body temperature.
8 Water balance.
9 Emotion.

Due to the connection with the central nervous system, stress and emotional disturbances may upset the hypothalamic–pituitary mechanism.

The pituitary gland

The pituitary gland (hypophysis) is known as the 'master' gland of the endocrine system owing to its influence on the other endocrine glands.

This gland is situated at the base of the brain in the sella turcica, or pituitary fossa, of the sphenoid bone. It is a small, round structure in the region of 13 mm in size and weighing approximately 0.6 g. A small stalk, called the infundibulum, connects the pituitary gland to the hypothalamus.

The pituitary gland is divided into two portions – anterior and posterior – which differ in origin, structure and function. The anterior section is composed of vascular, glandular tissue responsible for the production and secretion of six hormones. The posterior section is composed of nerve-like tissue supplied from the hypothalamus and is responsible for the storage and release of two hormones. Both sections are under hypothalamic control.

The anterior pituitary gland

The anterior pituitary gland (adenohypophysis) is connected to the hypothalamus by a network of blood vessels through which pass the releasing and inhibiting factors that control pituitary secretion. Pituitary trophic hormones exert their influence on other endocrine glands. Each trophic hormone influences a specific target gland. Hormones produced by the anterior pituitary are as follows.

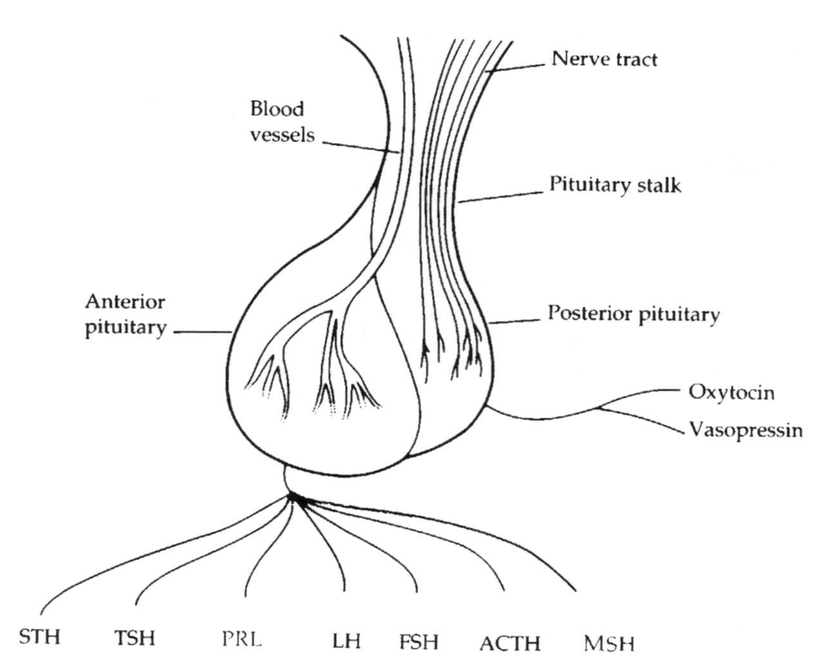

Figure 5.5 The pituitary glands

Somatotrophic (STH) or growth hormone

This controls the growth of bones and formation of tissues. It has an effect on protein, fat and carbohydrate metabolism and promotes the retention of nitrogen. Over-secretion of STH leads to pituitary giants in children or enlarged face, hands and feet in adults (acromegaly). Under-secretion in children results in pituitary dwarfs owing to under-development of the long bones.

Adrenocorticotrophin (ACTH)

This controls the adrenal cortex, stimulating secretion of steroids. Insufficient stimulation of adrenal cortex by ACTH results in the gland shrinking. Too much secretion causes enlargement of the gland and an excess production of androgens. Prolonged periods of stress result in increased stimulation of the adrenal gland and therefore higher levels of circulating androgens, which may result in increased hair growth.

Thyroid stimulating hormone (TSH)

This stimulates the thyroid to produce hormones for regulating body metabolism.

Gonadotrophic hormones

These stimulate the sex glands and are known as follicle-stimulating hormone (FSH) and luteinizing hormone (LH) in the female. LH in the male is known as interstitial cell-stimulating hormone.

FSH stimulates the ovarian follicle in the female to ripen and produce oestrogen. In the male the testes are stimulated to produce spermatozoa.

LH in the female stimulates the corpus luteum in the ovary and also secretion of oestrogen and progesterone. In the male, the testes are stimulated to produce testosterone. LH influences the development and maintenance of male sex characteristics. In both sexes LH stimulates androgen production by the sex organs.

Prolactin (PRL)

This is responsible for stimulating milk production from the breasts.

Melanocyte-stimulating hormone (MSH)

This is responsible for stimulation of melanocytes to produce melanin pigment, which results in darkening of the skin.

The posterior pituitary gland

The posterior pituitary gland stores two hormones which are produced by the hypothalamus. These are oxytocin and vasopressin.

Oxytocin

This stimulates contractions of the uterus during labour, and also aids the flow of milk after birth by stimulating lining cells of the breast ducts.

Vasopressin (ADH)

This is known as the antidiuretic hormone and is concerned with the maintenance of the body water balance. It influences the renal ducts of the kidneys and water metabolism.

The thyroid gland

The thyroid gland, the largest endocrine gland, consists of two oval lobes that are situated in the neck just below the larynx, either side and slightly to the front of the trachea. The two lobes are connected by the isthmus, which is a narrow part of the gland (see Figure 5.6).

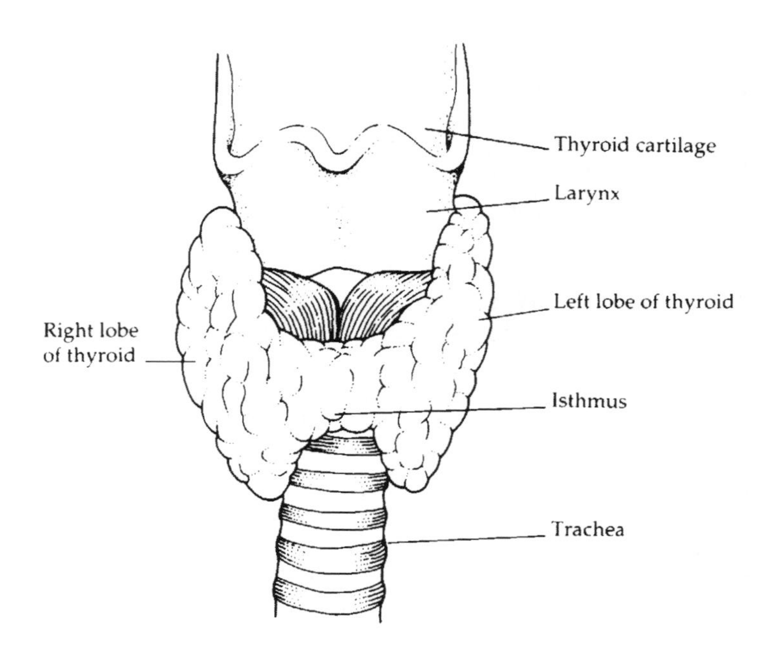

Figure 5.6 The thyroid gland

The thyroid relies on the normal functioning of the pituitary gland for stimulation by TSH to produce three hormones:

1 thyroxine
2 tri-iodothyronine
3 calcitonin.

These three hormones control metabolism, affecting general metabolism and function of the body. Calcitonin affects calcium metabolism. Iodine from food combines with protein to form thyroglobulin, which is later changed by the action of enzymes to thyroxine for release into the bloodstream as required.

Hyperthyroidism (Graves' disease)

When too much thyroxine is produced, a number of problems arise. The person becomes mentally full of energy but is unable to keep up physically. Because of increased metabolism, weight-loss occurs. Increased heartbeat, palpitations, profuse sweating, anxiety and heat intolerance are other symptoms associated with over-production of thyroid hormones.

Hypothyroidism (myxoedema)

When insufficient thyroxine is produced the body's metabolism will slow down. The individual may become slow and lethargic, lose concentration easily, feel the cold and experience steady weight gain. The body temperature is usually subnormal. The skin becomes dry, and hair growth may be sparse, dry and lifeless.

The parathyroid glands

The parathyroid glands are four small glands embedded on the back and side surfaces of the thyroid gland. Their role is to produce the hormone parathormone.

Parathormone and calcitonin are responsible for maintaining the calcium levels in the blood. Calcitonin is released when calcium levels are too low.

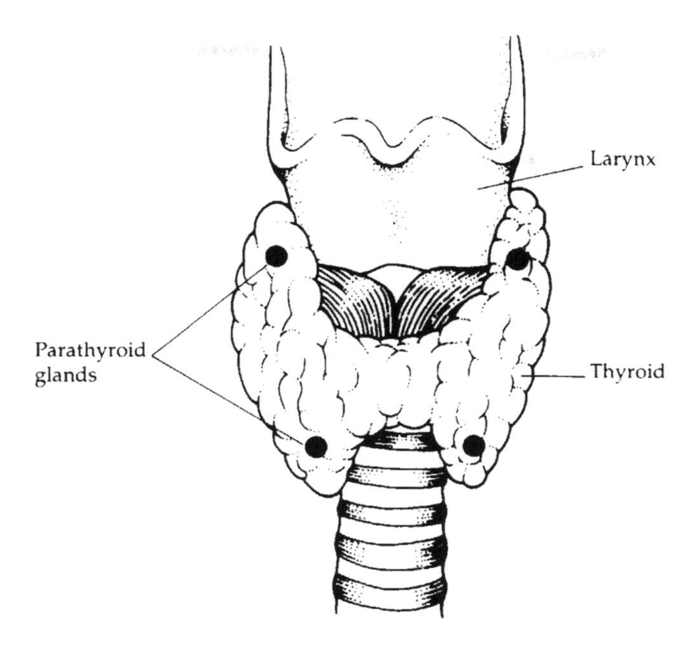

Figure 5.7 The parathyroid glands

When this happens the result is sparse, dry, brittle hair, which in turn, affects hair devlopment. High calcium levels can lead to serious bone disorders or kidney stones.

The pancreas gland

The pancreas is situated in the curve of the duodenum, behind the stomach. This gland has two functions:

1 Exocrine – production of pancreatic enzymes to aid digestion in the small intestine.
2 Endocrine – production of insulin and glucagon.

Specialized cells known as the islets of Langerhans form the endocrine section and are divided into alpha and beta cells. Alpha cells are responsible for the production of insulin, and beta cells for the production of glucagon.

Insulin controls carbohydrate metabolism. Its role is to lower blood sugar levels by promoting entry of sugar into the body cells for metabolism or by encouraging glycogen storage in the muscles and liver. Glucagon is responsible for raising the blood sugar levels by freeing stored glycogen

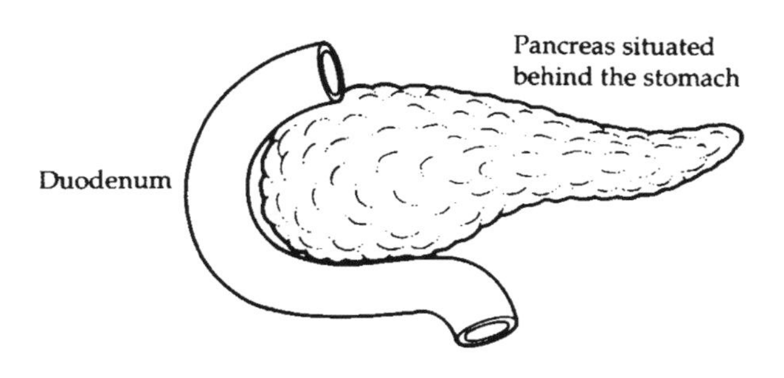

Figure 5.8 The pancreas

from the liver. *Diabetes mellitus* occurs when there is insufficient insulin present in the blood. This disorder affects both sexes but its development is not uncommon during pregnancy and menopause.

Symptoms associated with diabetes mellitus are:

1 Increased thirst.
2 Increased output of urine.
3 Weight loss.
4 Thin skin, with reduced healing ability.
5 Increased tendency to develop minor skin infections.
6 Lowered natural body defence against infection.
7 Decreased pain threshold when insulin levels are low.

The electrolysist should give careful consideration to the above points when planning treatment sessions for the client who has diabetes.

The adrenal glands

The two adrenal glands are found above each kidney. Each gland consists of an outer cortex, which is yellow in colour, and an inner reddish medulla.

The *medulla* is closely linked to the sympathetic nervous system and produces two hormones, adrenaline (epinephrine) and noradrenaline (norepinephrine). The medulla has a complex nerve supply which controls the dilatation and contraction of the blood vessels. The functions of *adrenaline* are as follows:

1 Stimulates metabolism resulting in the release of glycogen as glucose into the bloodstream thereby raising blood sugar.
2 Brings about the contraction of arterioles in the skin and internal organs.
3 Encourages increased oxygen intake by dilating the bronchioles in the lungs.
4 Dilates arterioles in the heart and muscles.

Adrenaline is responsible for the activation of the 'fight and flight' mechanism in times of fear, anger or emergency. When this mechanism is

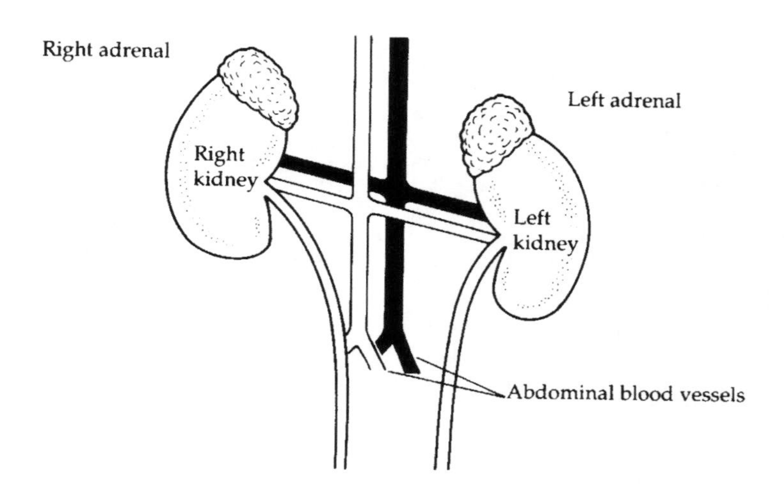

Figure 5.9 The adrenal glands

activated the muscles receive an increased supply of sugar, blood and oxygen and therefore function more efficiently. The heart rate is increased. The surface of the abdominal blood vessels contracts so the blood supply can be directed to the muscles. Movement of the digestive tract slows down.

Noradrenaline, the second hormone produced by the medulla, has an effect on the circulation by contracting blood vessels and raising blood pressure.

The *outer adrenal cortex* produces three groups of hormones:

1 Glucocorticoids.
2 Mineral corticoids.
3 Sex corticoids.

Glucocorticoids are concerned with carbohydrate, fat and protein metabolism. This group includes cortisone and hydrocortisone, which affect growth of connective tissue.

Mineral corticoids, which include aldosterone, regulate electrolyte (salt) and water-balance in the body.

Sex corticoids/steroids act as an auxiliary source of male and female sex hormones. These hormones affect the development and functioning of reproductive organs. They also influence the physical and temperament characteristics of both sexes.

When hydrocortisone levels fall, the hypothalamus produces more releasing factor, so stimulating the pituitary to produce more ACTH to restore hydrocortisone production by the adrenal cortex. As a result, androgen levels are raised, consequently increasing hair growth in androgen-sensitive follicles. Stress, hyperplasia and adrenal tumours may all bring about excessive steroid secretion, which in turn can stimulate hair growth.

The ovaries

The ovaries are two small glands which form part of the female reproductive system. They are situated in the pelvic cavity on each side of the uterus and are attached to the fallopian tubes by strong ligaments.

The ovaries have two functions:

1 to produce the hormones oestrogen and progesterone, and small quantities of androgens;
2 to produce ova (eggs).

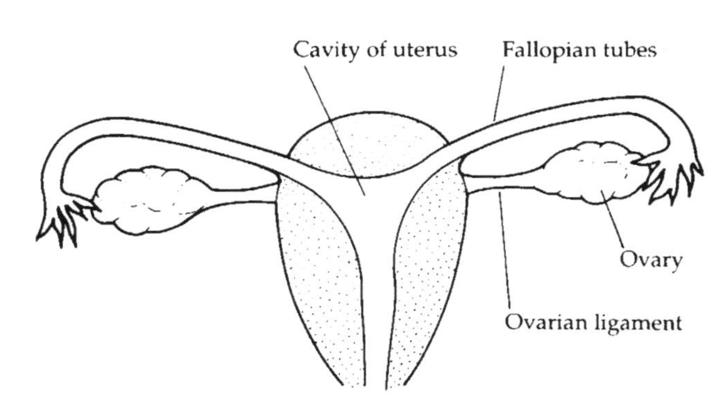

Figure 5.10 The ovaries

Ovarian hormones become active at puberty and are responsible for development of secondary sex characteristics. *Oestrogen* is concerned with breast development, growth of the milk ducts in the breast, health and growth of bones, and subcutaneous fat distribution. Oestrogen influences the menstrual cycle and thickens the uterus lining in preparation for conception. It also has an effect on the brain, skin, arteries, veins and muscles. Oestrogen production by the ovaries begins to decline as the menopause approaches. *Progesterone* is known as the pregnancy hormone. It is concerned with the development of the placenta, and maintenance of the pregnancy and also helps to prepare the mammary glands for lactation.

Menstrual disorders occur when the ovary fails to respond to stimulation by the pituitary gonadotrophic hormones or when there is an abnormal response to stimulation. Increased androgen production by the ovary may give rise to hirsutism.

The pineal and thymus glands

These are both endocrine glands; however neither gland has any known influence on hair growth. Recent studies have shown that the pineal has inhibitory and stimulatory influences on the hypothalamus (*The Cause and Management of Hirsuitism*, Greenblatt, Mahesh and Gambrel).

The pineal is a small gland situated at the base of the brain. The thymus lies high in the chest.

Review questions

1. Name the glands which make up the endocrine system.
2. What is a hormone?
3. Describe the role of the hypothalamus.
4. Name and give the functions of the pituitary secretions.
5. What is the role of thyroxine?
6. Describe the two functions of the pancreas.
7. List the symptoms associated with diabetes mellitus.
8. Describe the functions of the hormones secreted by the adrenal cortex.
9. What is meant by the 'fight and flight mechanism'?
10. Why is it necessary for the electrolysist to be familiar with the physiology of the endocrine system?

6 Hirsutism and hypertrichosis

The terms hirsutism and hypertrichosis are often confused, although there is a definite difference between the two. Both terms are used to describe excessive hair growth in women, but there the similarity ends. Figure 6.1 displays the main features of each.

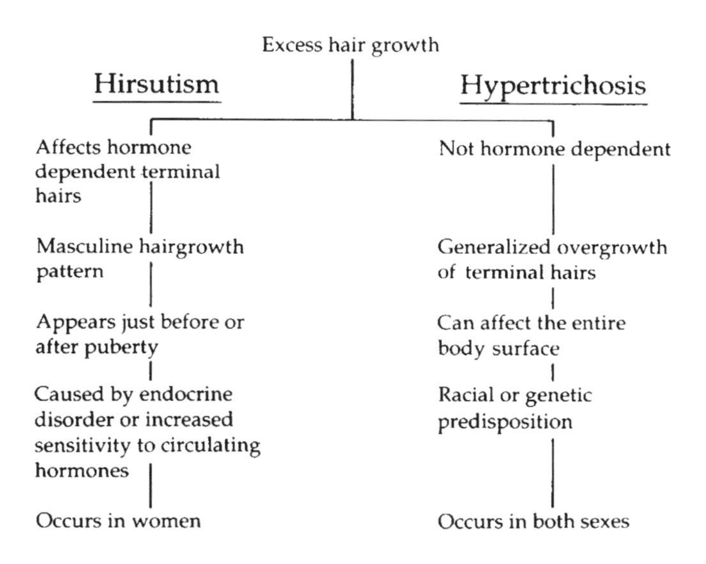

Figure 6.1 Excess hair growth

Hypertrichosis refers to a generalized over-growth of vellus and terminal hair in either sex. Hairs grow faster than normal, although there is no increase in diameter size. This type of growth is not due to a systemic disorder. Causes are genetic and racial tendencies or constitutional variation in follicle sensitivity.

Hirsutism refers to a masculine pattern of hair growth in women – one that is normal in men. There is an increase in cyclic growth, diameter of the hair and rate of growth. Hirsutism is considered a disease by many medical authorities and can arise from a serious underlying disorder that can often be diagnosed by some simple blood tests and detailed notes of medical history.

Hirsutism is caused by the following two factors:

1. Increased follicle sensitivity to normal levels of circulating androgens in the blood. This is referred to as primary hirsutism.
2. Increased androgen production by the adrenal glands and ovaries. This is referred to as secondary or true hirsutism.

The onset of primary hirsutism occurs at puberty, increasing until the thirties when it stabilizes; secondary hirsutism begins just before or just after puberty. Secondary hirsutism is due to an endocrine disorder that causes increased hormonal secretion by the glands.

The total number of hair follicles for an adult human is estimated at 5 million, with 1 million on the head of which 10 000 alone cover the scalp. The number and distribution of hair follicles is the same in both sexes. It is genetic predisposition which determines the sex of an individual and specific hair growth patterns. Hormones are responsible for influencing and stimulating growth at the follicle cells, which in turn determines the pattern, quantity, texture and distribution of body hair.

It is important for clients to understand that both primary and secondary hirsutism can be managed successfully with medical treatment, which will correct or control endocrine disorders, thereby preventing the development of new growth. Electrical epilation will eventually remove the existing hair growth permanently. It must be explained to the client that established growth did not appear overnight, and that time is therefore needed to eliminate the problem permanently. The fact that results are slow but sure should be emphasized.

Causes of hair growth

To achieve a successful result with electro-epilation it is necessary for the electrolysist to understand the causes of hair growth. This cannot be fully achieved without having first studied the endocrine system (see Chapter 5).

Consideration should be given to genetic/heredity tendencies, topical causes and sensitivity of follicles to circulating androgens in the bloodstream.

The highest percentage of hirsute women have either a hereditary predisposition or some subtle change in their androgen production. Hair distribution, type of hair and rate of growth varies from one race to another and from one person to another, for example, Caucasians have a higher number of hair follicles than the Japanese and Chinese. Some find the presence of hair both desirable and acceptable, whereas others prefer a complete absence of hair.

Sensitivity of follicle end receptors to circulating androgens encompasses a number of factors, from topical friction to endocrine influences. The degree of hormone sensitivity can vary from one area to another and between different individuals. Increased sensitivity of androgen receptors in the pilo-sebaceous unit may be a dominant or recessive hereditary trait.

Topical causes include sustained friction, e.g. a plaster cast. Friction causes an increase in local blood circulation to the skin. This type of hair growth is temporary and usually returns to normal shortly after the plaster cast has been removed.

The tweezing of individual hairs tears out the lower follicle. The reconstructed follicle is usually stronger, with a better blood supply. Vellus and fine hairs are frequently removed at the same time, so aggravating the condition. Waxing of facial and vellus hair will have the same effect as tweezing. Hairs that have been waxed or tweezed invariably leave behind distorted follicles, so hindering electro-epilation unless blend or galvanic electrolysis is used.

Endocrine influences can be divided into normal systemic and abnormal systemic conditions. Normal systemic conditions include puberty, pregnancy, menopause, stress and sensitivity of follicle end organ receptors to androgens. Abnormal endocrine conditions include polycystic ovary syndrome (Stein–Leventhal syndrome), Cushing's syndrome, adrenal tumours, ovarian tumours, diabetes mellitus and anorexia nervosa.

Androgens and growth hormones increase the size and diameter of hair growth. Cortisols, oestrogens and thyroid hormones alter and influence without increasing hair growth. Although oestrogens do not initiate growth they can prolong an existing hair growth cycle (*The Cause and Management of Hirsutism,* Greenblatt *et al.*).

Surgery can also contribute towards the development of unwanted hair, e.g. total hysterectomy, which involves the removal of the ovaries and therefore alters the hormone balance.

Normal endocrine influences

Puberty

The anterior pituitary gland secretes gonadotrophic hormones, which influence the target organs. These are ovaries in the female and testes in the male. At this stage of life the adrenal cortex becomes active. Between them, the adrenal cortex and the gonads secrete large quantities of steroid hormones into the circulatory system.

The appearance of both pubic and axillary hair is due to the increased level of adrenocortical androgens. Hereditary sensitivity in combination with the amount of hormone produced is responsible for the appearance of hair in other areas. Women produce smaller quantities of androgen than men. Androgen levels increase in women during puberty, pregnancy and menopause.

Pregnancy

During pregnancy there is an increase in hormone activity. On occasions, excess androgens are produced, with the result that fine hair growth may appear on the lip, chin and sides of the face. Quite often this is a temporary growth that disappears shortly after the end of pregnancy.

Menopause

The menopause marks the end of a woman's reproductive life. Between the ages of 40 and 50 there is a gradual decline in the oestrogen and progesterone levels owing to ovarian tissue slowly ceasing to respond to stimulation by the gonadotrophic hormones of the anterior pituitary gland. Facial hair often develops because of the increased level of circulating androgens.

Emotional stress

The adrenal glands could be termed the 'stress glands' of the body. They control the 'fight and flight' mechanism. When the body is under stress, either emotionally or physically, the activity of the adrenal glands is increased. During stress the hypothalamus triggers the anterior pituitary gland to produce increased levels of adrenocorticotrophic hormone (ACTH). This in turn stimulates the adrenal glands to produce adrenaline, at the same time increasing androgen production. When stimulation takes place over a prolonged period, hair growth may occur.

Medications

Certain medications are known to stimulate or aggravate hair growth. Hormonal medications with androgenic properties may cause hair to grow in a masculine pattern. This type of drug includes testosterone, some oral contraceptives, and anabolic steroids.

Endocrine disorders that affect hair growth

Endocrine disorders arise out of a glandular defect that may be inherited from either parent or may be an acquired disease. Disorders that affect hair growth are: Cushing's syndrome, Stein–Leventhal syndrome and adrenal neoplasms.

Adrenal influences in hirsute women

People vary greatly in their skin sensitivity to androgens owing to their genetic background: a high level of circulating androgens may have no effect on one individual, whereas a low level may induce hair growth in another.

Adrenogenital syndrome

Adrenogenital syndrome fortunately is a rare condition and very unlikely to be seen by the practising electrolysist. The effect of androgen disorder is more noticeable in girls, who show an abnormal development of the external genitalia. In extreme cases, the clitoris protrudes and may be mistaken for a penis. The adult female will eventually have a male build and develop a deep voice as well as a male distribution of hair growth.

When a boy is affected, he will reach puberty between the ages of three and five years, with secondary sex characteristics becoming noticeable. High androgen levels in both sexes cause rapid body growth that stops early. The epiphyses in the bone fuse at an earlier age than normal.

Symptoms, which may be present in the adult woman, are due to oversecretion of androgens. These include receding hairline, the appearance of bald patches, increased hair growth in a masculine pattern on the face and limbs, breasts becoming atrophied, and the menstrual cycle may be absent or become irregular. Feminine fat is replaced by masculine muscle.

Adrenal tumours

These are small, non-encapsulated masses and are usually single, solitary nodules. Adrenal tumours (also known as neoplasms) can occur at any age, although they are more common around the age of 30–40 years. Their presence is indicated by a sudden onset of virilization or Cushing's syndrome. This type of tumour may be masculinizing, causing hirsutism, increased muscle mass and deepening of the voice in women.

Virilizing congenital adrenal hyperplasias

Virilizing congenital adrenal hyperplasias are the result of enzyme deficiency that affects the production and levels of adrenal hormones. Partial or total enzyme deficiency results in decreased cortisol levels. At the same time production of androgen is increased by the adrenal cortex, leading to an excess and thereby causing hirsutism.

Cushing's syndrome

Cushing's syndrome is brought on by an excess of glucocorticoids as a result of a tumour or excessive adrenal cortex function. Pituitary tumours and certain cancers, such as lung and pancreatic cancers, result in overproduction of ACTH. Increased levels of ACTH lead to an excess of cortisol, androgen and aldosterone production.

There are a number of symptoms associated with Cushing's syndrome, which include: osteoporosis owing to decreased calcium absorption; obesity of the trunk with purple stretch marks on the abdomen; muscular weakness and wasting of the limbs; the skin becomes thin and bruises easily; rounding of the face due to fat deposits in the cheeks; diabetes can occur due to increased steroid production; salt retention leading to high blood pressure and oedema; vellus hair growth due to increased cortisol levels; and excess androgens possibly leading to hirsutism.

Stein–Leventhal syndrome

This condition is also known as polycystic ovary syndrome. Polycystic ovaries are capable of secreting large quantities of androgens. Symptoms include enlarged ovaries with numerous follicular cysts, irregular menstrual cycle, weight gain and the development of excess hair growth. The onset of hirsutism is gradual, occurring from puberty onwards. The electrolysist may be the first person to observe the signs that indicate a client may be suffering from this condition. Referral of the client back to her doctor for further investigation is advisable.

Masculinizing ovarian tumours

This type of tumour is rare, but when present it may cause an excess production of androgen. Hirsutism caused by masculinizing ovarian tumours has a rapid onset, usually but not always later in life. Menstruation may stop and will only start again after surgery.

Archard–Thiers syndrome

This is another relatively rare endocrine disorder. The characteristics include: generalized obesity, diabetes mellitus, hypertension and hirsutism. Hair growth is most noticeable on the face, with emphasis on the moustache and beard. Menstrual disorders may occur.

Anorexia nervosa

Anorexia nervosa is a condition that usually affects adolescent girls. It is a condition that involves the nervous, digestive and endocrine systems. Anorexia nervosa exhibits a number of characteristics, which include an increased growth of vellus hair on the face, trunk and arms, a persistent refusal to eat food, the absence of menstrual periods, weight loss and wasting muscles, and emotional stress or disturbance.

Other features associated with this condition are the sufferer's disgust with her personal appearance, a strong conviction that she is overweight and a lack of self-esteem. Initially it is possible to keep the problem hidden from other people, but over time a sharp-eyed relative, friend, associate or tutor will become aware of the situation. The anorexic person will persistently refuse to acknowledge that a problem exists.

Evaluating the cause of hair growth

Before electro-epilation treatment begins the cause of the hair growth should be assessed. It may well be advisable for the client to be referred for medical investigation, after which electro-epilation can commence.

A detailed consultation may give many clues as to the cause of the problem. The hair growth rate and pattern should be looked at. Questions should be asked relating to the onset of the problem. Hair growth that appeared at puberty and has not altered since gives little cause for concern, usually responding well to electro-epilation. A recent onset of hair growth, or one that is getting worse, needs further investigation. Examination of the skin and questions relating to the client's weight may also give valuable clues, for example, the presence of acne-type lesions, excessive sebaceous secretion and an increase in weight may be connected to excess androgen production.

Review questions

1. (a) What is meant by the terms 'hirsutism' and 'hypertrichosis'?
   (b) Compare the two conditions.
2. Explain the role of androgens in hirsutism.
3. Explain how emotional stress can stimulate hair growth.
4. What is the cause of virilizing congenital adrenal hyperplasia?
5. Describe the signs and symptoms of Stein–Leventhal syndrome.
6. Name the characteristics and symptoms associated with anorexia nervosa.
7. How does a detailed consultation help when evaluating the cause of hair growth?

7 Natural hormone changes in a woman's life

Throughout a woman's reproductive life her hormone balance and levels are constantly changing. From puberty through to menopause, the levels of oestrogen and progesterone are never static. The different phases to consider are:

- Puberty.
- The menstrual cycle.
- Pregnancy.
- The menopause.

Each relies on the smooth functioning of the feedback mechanism between the hypothalamus, the anterior pituitary and the ovaries. When any one of these three fails to function at 100% efficiency, a hormone imbalance will occur that may result in disorders such as endometriosis, polycystic ovaries or menstrual irregularities.

Puberty

Puberty marks the time of change from childhood to womanhood. Hormone levels start to rise, sex organs become functional and the menstrual cycle begins. Hormonal changes precede physical changes and begin well before the first menstrual period.

Puberty normally begins around the age of 11 years but can vary between the ages of 10 and 15 years. The onset of puberty may be influenced by genetic, endocrine, nutritional, physical and environmental factors.

This process begins when the hypothalamus stimulates the anterior pituitary gland to secrete gonadotrophic hormones that activate the sex glands. The hormones secreted by the anterior pituitary gland at this time are:

- *Adrenocorticotrophic hormone*, which stimulates the adrenal cortex to produce oestrogen and progesterone in the female, testosterone and a little oestrogen in the male, also androgens in both sexes.
- *Follicle stimulating hormone*, which influences the ovaries to stimulate the maturing Graafian follicles and produce oestrogen.
- *Luteinizing hormone*, which stimulates the formation and secretion of the corpus luteum, which in turn secretes progesterone.

Changes that take place during puberty

Puberty is not an easy time for the young adolescent girl or for her parents. There are a number of psychological and physical changes. The appearance of acne can affect self-confidence. The individual can become difficult, argumentative, shy or aggressive and may require a great deal of patience.

Initially, the menstrual cycle is irregular but usually settles down in time. Fine, dark hair growth may appear on the upper lip and sides of the face due to specific sensitivity to circulating androgens. The rate at which the individual develops in comparison to her contemporaries often causes stress.

Normal changes that take place are as follows:

- Breast development.
- Alteration of body contours due to the laying down of feminine fat deposits, widening of the pelvis, strengthening of the muscles, and change in the shape of the face and jaw.
- Growth of pubic and axillary hair.
- Maturation of the genital tract.
- Increased growth in height and development.
- Emotional and temperament changes.
- Start of menstruation – the average age being 12 years 8 months.
- Adrenal cortex reaches maturity due to stimulation by the hypothalamus and anterior pituitary glands.

The menstrual cycle

The normal menstrual cycle starts at puberty and usually finishes during the menopause, with a natural interruption during pregnancy. Conditions such as endocrine disorders, stress, anorexia nervosa and surgery such as hysterectomy can alter, stop or suppress the menstrual cycle. The average cycle is 28 days but can vary between 17 and 42 days (see Figure 7.1). It is the pre-ovulatory phase that varies in length, with the post-ovulatory phase lasting 14 days.

The cycle begins when the pituitary gland secretes follicle-stimulating hormone which induces ripening of an ovarian follicle into a mature Graafian follicle. The follicle grows and expands, finally appearing on the ovary surface. Oestrogen, secreted by the growing follicle, thickens the uterus lining (endometrium). Glandular size and richness of blood vessels are increased. After approximately 10 days the pituitary gland secretes luteinizing hormone which, in combination with follicle-stimulating hormone, causes the Graafian follicle to rupture and release an ovum (egg).

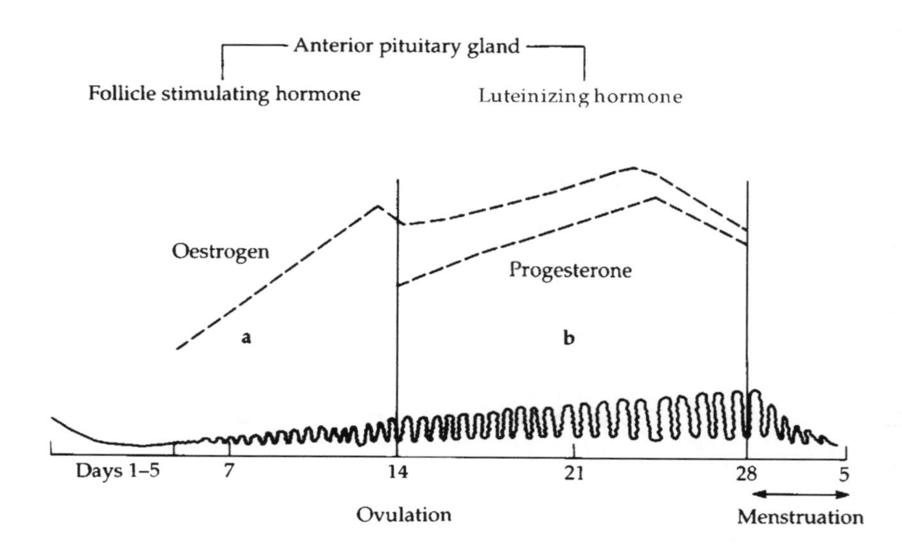

Figure 7.1 The menstrual cycle

The egg passes down the fallopian tube into the uterus, ready for fertilization to take place. The level of follicle-stimulating hormone falls at the time of ovulation and remains low for the rest of the cycle.

When the egg breaks free from the ovary it leaves behind a small amount of scar tissue. The tissue develops into the corpus luteum – a yellow gland. For approximately two weeks the corpus luteum produces progesterone to enrich the lining membrane of the uterus in preparation for pregnancy. The blood supply to the uterus is increased. When fertilization does not take place after 12–14 days, the corpus luteum shrinks, progesterone secretion stops, oestrogen levels fall, and blood vessels close and die. Menstruation will then take place. The lining membrane (endometrium) breaks down and is shed in the form of blood, together with the unfertilized egg. Menstruation lasts between 3 and 5 days.

Menstrual cycle ending in pregnancy

The egg becomes implanted into the hormonally prepared uterus lining when it is fertilized by a sperm. A foetus will then grow. The ordinary ovarian cycle is suspended during pregnancy, which usually lasts 9 months (39 weeks) from the date of conception.

After fertilization of the egg the corpus luteum continues to grow, reaching its peak after approximately 6 weeks. The activity of the corpus luteum falls off after approximately 2 months, finally stopping around the fourth month. The role of the corpus luteum during the early stages of pregnancy is to secrete large amounts of progesterone, which is essential for development of the placenta.

The placenta, which develops from the implanted embryo, also produces progesterone, gradually taking this role over from the corpus luteum. The purpose of placental progesterone is to maintain pregnancy and help prepare the mammary glands for lactation.

Some women notice an increase in excess hair growth during pregnancy; also the appearance of telangiectasia and spider naevi. Often this is temporary and will disappear without treatment shortly after the pregnancy has ended. Telangiectasia appearing on the legs during pregnancy frequently remain, but treatment by short-wave diathermy for these particular capillaries is not usually successful.

Premenstrual tension

Premenstrual tension (PMT) is the name given to a collection of mental and physical symptoms which occur for up to ten days before menstruation. Symptoms include fluid retention, temporary weight gain, distended abdomen, swollen ankles, breasts becoming tender, irritability and temperament changes. Accidents and clumsiness are common at this time. Pain threshold is lower, therefore the individual is often more sensitive to current intensity during electro-epilation. Hair growth appears to be faster immediately prior to menstruation.

The cause of these symptoms is believed to be an imbalance between oestrogen and progesterone.

The menopause

It is possible that a woman will live a third of her life after the menopause. The menopause is the time when a woman's reproductive function comes to an end. Menstruation becomes irregular and finally stops. The ovaries' supply of eggs dries up. There is a decline in hormone production by the ovaries from the age of 40 years onwards, with the menopause usually occurring between the ages of 45 and 55 years. Some 90% of women stop menstruation between the ages of 44 and 55 years. As the level of oestrogen decreases the amount of free testosterone increases.

A premature menopause will be brought about by surgical removal of the womb and ovaries, i.e. a total hysterectomy, or by treatment with chemotherapy.

There are a number of symptoms a woman may experience when she reaches the menopause. Unwanted hair growth may appear due to the lowering of oestrogen levels and follicle sensitivity to circulating androgens. There may be varying degrees of vasomotor instability, giving rise to night sweats and hot flushes. Depression, headaches, bladder irritation, vaginal dryness, painful intercourse, cystitis, insomnia, loss of confidence, poor concentration and tiredness as well as demineralization of bones (osteoporosis) and aching joints can all occur at this time of life.

Intermediate symptoms are dry hair and skin, dryness of the vagina, loss of libido and bladder infections. Long-term loss of oestrogen may result in osteoporosis, heart attacks or strokes. Oestrogen has been found to reduce strokes by 30% and heart attacks by 50%.

Psychologically a woman may feel that her useful life is over. If children have left home she may suffer from what is commonly referred to as 'the empty nest syndrome', whereby she feels that she is no longer needed.

In fact, when viewed from a positive angle, the menopause marks a new and interesting phase in life. There is freedom from a number of past responsibilities. Demands on finances may be considerably less. There is freedom from the risk of pregnancy. Quite often there is more time for partners to enjoy each other's company.

Hormone replacement therapy (HRT)

It is possible to obtain relief from many of the symptoms associated with the menopause through hormone replacement therapy. It is not a panacea for every woman, nor is it the fountain of eternal youth as some would have women believe. However, medical research has found that it has a number of advantages. In geriatric women, it is increasingly recognized that hormone replacement therapy can preserve muscle tone associated with the bladder, therefore helping to prevent incontinence. The risk of heart attacks and strokes appears to be reduced. Additional benefits of hormone replacement therapy are an increased sense of well-being through relief of physical symptoms, prevention against loss of bone (osteoporosis) and also improved memory and possible delayed onset or lowered rate of Alzheimer's disease.

Progesterones are added to oestrogen for two main reasons:

1. To prevent endometrial hyperplasia and its progression to carcinoma. Progesterones oppose the mitotic or proliferative effects of oestrogen, which may lead to endometrial hyperplasia.
2. In sequential therapy to promote a regular and predictable bleed. Progesterone induces secretory transformation, a process that activates

enzymes involved in oestrodial metabolism and endometrial shedding when progesterones are withdrawn. This ensures that menstruation occurs which will bring about the removal of potentially neoplastic cells.

Progesterone may be used in HRT in three ways:

1 As a cyclical monthly course in combination with oestrogen to produce a monthly bleed and protect the endometrium.
2 Continuously with oestrogen – so protecting the endometrium without a monthly bleed.
3 Progesterone-only therapy in women who are contraindicated to oestrogen due to a history of conditions such as breast or endometrial cancer.

There are a number of side effects with oestrogen, namely breast tenderness, nipple sensitivity, nausea, leg cramps and epigastric discomfort. A few women gain weight. These side effects can be alleviated with the use of combined oestrogen and progesterone preparations.

Dr Robert Greenblatt, an eminent endocrinologist in the United States, believes that the administration of testosterone along with oestrogen enhances the benefit of hormone replacement therapy and frequently eliminates depression. Energy levels and sex drive are improved. There is relief from migraine-type headaches and also considerable improvement in any arthritic condition. Dr Greenblatt also believes that the use of progesterone not only guards against hyperplasia but also against painful breasts, weight gain and irregular bleeding which may occur when using oestrogen only (*No Change*, Wendy Cooper, 1990).

In a lecture given to members of the Institute of Electrolysis, Dr John Studd, consultant gynaecologist at King's Hospital, London, spoke on the use of oestrogen and testosterone in the fight against osteoporosis. There is growing evidence that the use of hormone replacement therapy in the combined form can actively encourage the laying down of new bone. Osteoporosis and its associated complications cost the National Health Service many thousands of pounds each year. It is possible that this cost could be reduced considerably by the use of HRT.

There are a number of other benefits to be obtained from the use of hormone replacement therapy. These include improved muscle tone, firm breasts, supple joints, prevention of dry, wrinkled skin, also healthy hair and nails. Oestrogen helps to maintain the suppleness and elasticity of the vagina, so eliminating painful intercourse.

The main difference between the contraceptive pill and hormone replacement therapy is that the pill uses artificial hormones, whereas hormone replacement therapy mainly uses natural hormones. The contraceptive pill increases hormone levels, whereas hormone replacement therapy replaces the decreased hormone levels.

Hormone replacement therapy does not claim to prevent the ageing process or provide instant rejuvenation. What it does is to slow the whole procedure down, thereby allowing a woman to grow older gracefully and gently.

Therapy may be given in five ways:

1 Orally in the form of pills.
2 Transdermal patches.
3 Implants.
4 Vaginal creams.
5 Pessaries.

Orally

Hormone replacement therapy may be prescribed for oral use in the form of Premarin, which is oestrogen only, or as a combination of oestrogen and progesterone in the form of Cycloprogynova and Prempack-C. The use of progesterone over a period of 10–12 days ensures that any build-up of the endometrium is shed on a regular basis in the same manner as the natural cycle. Oestrogen only is most commonly used for those women who no longer have a uterus. Oral hormone replacement therapy can be varied to meet the needs of the individual.

A dose of oral oestrogens needs to be substantially higher than a non-oral dose. This is because oral oestrogen travels via the portal vein to the liver. A certain percentage is lost in the digestive tract.

Transdermal patches

Transdermal patches are small, transparent patches impregnated with either oestrogen only or combined cyclical HRT, that is, oestrogen and progesterone. Transdermal patches deliver a constant amount of hormones through the skin. They are usually worn on the lower abdomen, the buttock or the thigh. These patches are replaced at regular intervals (as prescribed by the medical practitioner). The advantage of patches is that the hormones are absorbed directly into the bloodstream, thereby avoiding loss through the digestive tract. The disadvantages are that the patch can easily be dislodged and skin irritation may occur. Women who have not had a hysterectomy must use the combined oestrogen and progesterone patches.

Skin reactions occur at the patch site in up to 30% of women. These can be minimized by applying the patch to the buttock, which is less prone to skin irritation; also, patches can be applied in a different position every 24–48 hours. Patches that contain alcohol should be left to stand for 10–15 seconds with the backing strip removed before application to the skin. Hot, humid climates increase the possibility of skin reaction.

Implants

The hormone implant is a painless, minor surgical procedure. The lower abdomen is anaesthetized. A small incision is made and a metal tube inserted. The appropriate hormone pellets are implanted into the abdominal fat through the tube, then the tube is removed and the incision stitched. The effect of the pellets lasts for up to 6 months. Some practitioners favour the use of testosterone pellets in addition to oestrogen pellets. It is felt that testosterone may well enhance the benefits of oestrogen. The use of testosterone may cause increased hair growth.

Vaginal creams

A number of topical oestrogen preparations are available for short-term relief of dryness and atrophy of the vagina in women who lack systemic symptoms of oestrogen deficiency or who do not wish to go on systemic treatment.

Review questions

1. Name the different phases in a woman's life when major hormonal changes occur.
2. Describe the changes that take place at puberty.
3. Give a detailed description of the menstrual cycle.
4. What is meant by the term 'premenstrual syndrome'?
5. Explain the benefits that may be obtained from hormone replacement therapy.
6. Name the different ways in which hormone replacement therapy can be administered.

8 Gender reassignment

A transsexual is a person who feels very strongly that he/she is trapped in the wrong biological body. The individual is firmly convinced that he/she has the wrong physical characteristics for his/her true sex.

There are two types of transsexuals. Male to female transsexuals have a male body and a female mind, whereas female to male transsexuals have a male mind in a female body. It is the male transsexual that the electrolysist will encounter.

Some people think that transsexualism is a psychological condition, with no physical cause. Recent research into the human brain has shown that male and female brains are different in structure. Scientists who have studied the brains of transsexuals have found that their brains are structured like those of their psychological rather than their physical sex (*The Looking Glass Society*, www.looking-glass.greenend.org.uk).

Feelings experienced by transsexuals usually occur at an early stage in life, often before the age of five years. Some are aware they want to be the other sex but do not always understand why; others are well into middle age before they understand the truth about themselves. This is not a desire, but a desperate need to live and function as the opposite sex.

Transsexuals should not be confused with transvestites. 'Transvestite' is the term applied to men who cross-dress in women's clothes *from time to time for emotional or sexual pleasure*, but who have no desire to live and function as women. Neither should a transsexual be referred to as homosexual: there are fundamental differences between the two. *Homosexuals enjoy their own sex and prefer to relate to their own gender sexually.* A homosexual man has no desire to undergo surgery, or to live as a woman.

Once experienced, transsexual feelings grow stronger as time goes by. Individuals find it difficult or impossible to suppress these feelings, either through their own efforts or by medical intervention, that may include psychoanalysis, psychotherapy, electric shock or the administration of drugs.

There are two constructive courses of action open to these people:

1 to try to adjust to living as best (s)he can as his/her biological sex; or
2 to seek gender reassignment (sex change) by surgery.

The decision to go through surgery is not an easy one. There are many emotional, social and physical readjustments to be made, which can take place over a considerable period of time. Several barriers have to be overcome before surgery takes place, and a number of adjustments in life have to be made afterwards. The true transsexual must be able to pass successfully as a member of the opposite sex.

Male to female gender reassignment

There are a number of conditions that must be met before surgery can take place. The person concerned must:

1 undergo considerable stringent psychiatric assessment to establish that they are genuinely *gender dysphoric* (transsexual);

2 be single;
3 be able to live continuously and be self-supporting in their chosen role for a specified period of time, approximately two years;
4 be able to form social relationships as a woman with both sexes;
5 be emotionally stable.

Male to female surgery

Gender reassignment of male to female involves several hours of surgery. It is a long, slow procedure that can cause many problems unless carried out by a skilful surgeon.

The surgery involves:

1 Removal of the testes.
2 Building a pseudo-vagina.
3 Breast enlargement.
4 Reduction of the Adam's apple when necessary.

The process is painful and postoperative convalescence may take up to 6–8 weeks.

Surgery for gender reassignment is available either privately or through the National Health Service. However, surgery for breast augmentation, nose reduction and reduction of the Adam's apple are considered to be cosmetic surgery and are not usually available through the National Health Service. It is interesting to note that although the operation is available through the National Health Service in Britain, it is not legally possible to alter the sex stated on the birth certificate, and national insurance records cannot be altered, therefore surgery will have no effect on national insurance contributions or future rights to benefits and pensions.

Hormone treatment

For two years prior to surgery a programme of hormone treatment will be prescribed. This involves taking oestrogen orally, and possibly an androgen suppressant. Premarin is the usual form of oestrogen prescribed, but in a much larger dosage than that for the contraceptive pill or hormone replacement therapy. The dosage is usually reduced after surgery.

Physiological effects during hormone therapy

A number of physiological changes take place when hormone therapy is taken over a prolonged period. These include:

1 Development of small breasts, with pigmentation and enlargement of the areola.
2 Increase in, and redistribution of, fat cells, which affects the contours of the body.
3 Improved condition of scalp hair.
4 Improvement in skin tone.
5 Softening of the beard, and hair growth may occur.
6 Decrease in the size of the testes may occur.
7 Long-term impotence and sterility may occur.

Oestrogen will not:

1 Cause the masculine voice to become more feminine – this can only be helped by speech therapy.

2 Decrease or alter beard and body hair growth.
3 Reverse a receding hair line or cure baldness.

Electro-epilation

The presence of a beard poses many problems for the male to female transsexual who is living and working as a woman. Psychologically, it is better to have successfully removed the beard before the operation takes place. To have to continue shaving after surgery has been completed is demoralizing and often leads to severe depression. Adjusting to a new body image after surgery causes enough difficulties without the constant reminder of the preoperative life.

Electro-epilation will successfully eliminate the beard – but it is time consuming. It is a long, slow process influenced by the density and strength of the hair growth, sensitivity of the skin and pain threshold of the individual. The minimum length of time for treatment to a medium beard growth will be at least 2–3 years, and longer for a strong growth. Duration of treatment will need to be at least 2 hours a week. Treatment is rarely available on the National Health Service, but for the persistent individual it is sometimes possible.

Electro-epilation plan for the transsexual client

It must be remembered that treatment for these clients will be a long and slow procedure. As previously mentioned, it is better to start treatment at the earliest opportunity before surgery, to clear as much hair growth as possible prior to gender reassignment. Treatment time span may be up to 18 months or 2 years before surgery, and up to a further 3 years after surgery. Two hours a week is advisable, with the time decreasing as progress is made.

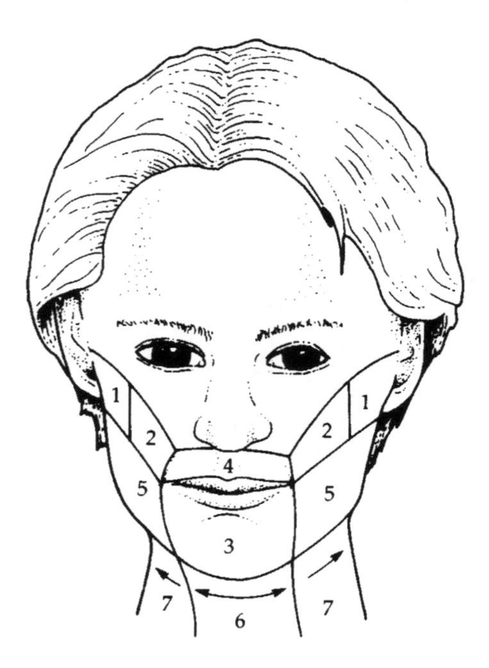

Figure 8.1 Treatment procedure

Treatment procedure

1 Starting at the ears, gradually work downward on sideburns and sides of face to thin out the hair.
2 Work downwards on cheeks. At this stage it is better to ignore regrowth, and concentrate on fining down the texture in order to tone down the blue shadow.
3 Proceed to the chin and upper lip. Care must be taken to avoid overtreatment of the upper lip. By this stage the electrolysist should be able to gauge the skin's response to treatment and work accordingly.
4 Progress to the neck and jaw line, gradually moving down to the sternum and breast area.

It will still be necessary for the transsexual client to shave between treatments. Psychologically, the sooner the hair growth can be diminished the better (remember, that these clients will be living and working in their female roles for some considerable time before surgery takes place, so the presence of a beard is both embarrassing and socially unacceptable).

When treatment is well under way, continue with 2 hours a week, with one hour spent concentrating on new growth and one hour on reducing regrowth.

With *careful planning and conscientious work* the skin will not be harmed. Method of treatment depends on the skin's reaction. The author has found the blend method to be the most effective treatment for these clients. Owing to the nature of the hair growth, the use of more galvanic than normal may be necessary, i.e. up to 1 milliamp , and therefore the use of cataphoresis after treatment is advisable in order to neutralise the surface effect of sodium hydroxide. When treating with short-wave diathermy there is a risk of pigmentation due to the high levels of oestrogen being prescribed.

The importance of correct home and after-care routine must be stressed. Advice on skin-care routine and products to improve the condition of the skin is often necessary. The electrolysist will probably become more of a confidante than with any other client, and can help with tips on make-up and leg waxing etc. When the transsexual client is living and working in her new role, the electrolysist is often the only person with whom she can discuss her problems and difficulties.

It can be seen from the preceding text that gender reassignment is not an easy option. The procedure involves a complete change of lifestyle, which may mean loss of family and friends; also, a change of career or job may be necessary. The decision requires courage and a great deal of determination.

Review questions

1 Explain the difference between a transsexual, transvestite and a homosexual.
2 What are the conditions that must be met before male to female reassignment can take place?
3 Describe the hormone treatment which is necessary prior to surgery.
4 State the physiological effects which occur during hormone therapy.
5 Give a treatment plan for the transsexual client.
6 Explain the complications that may arise owing to the amount of treatment required by a transsexual client.

9 Needles

The correct choice of needle plays an important role in the successful destruction of the lower follicle and dermal papilla. The function of the needle is twofold: first it acts as a probe, sliding smoothly and easily into the follicle; secondly it acts as an electrode, allowing the current to flow evenly to the tip of the needle. A good quality needle should have a smooth, polished surface with a well-defined, yet rounded tip.

How often do operators give a thought to the development and quality of the needle they are using or the effect the needle has on the efficiency of the treatment, not to mention the comfort of the client? The history of the epilation needle is fascinating, and much of the early information available to electrolysists to date can be attributed to meticulous and thorough research undertaken by Derek Copperthwaite, Editor of *International Hair Route Magazine*, Canada.

It is recorded that when Dr Charles Michel conducted his first experiments in hair removal using galvanic current, he used a fine, no. 8 sewing needle. The original needles used by Dr Michel were in all probability produced by one of the many needle factories to be found in and around the Redditch area in England. The demand for epilation needles in the early days was very small. The American distributors of medical equipment looked towards fine sewing needles that were being imported from Europe. At that time the demand for electrolysis needles was not sufficient to justify the expense involved in producing a specialized needle. It was found that a fine sewing needle could be adapted by either leaving out the eye or cutting it off altogether. Needle holders were made from wood or hard rubber.

In the late 1880s Daniel Mahler (founder of the Instantron Company, Rhode Island) was using fine English darning needles similar to those used by Dr Michel. Some 50 years were to pass before there was sufficient demand to warrant the manufacture of a specialized electrolysis needle. Initially, these needles were produced from sewing needles, manufactured in Redditch, England. The diameter was approximately one-hundredth of an inch, the length being in the region of 2.5 inches. Needles were buffed and polished by hand, usually four at a time, on a jewellery-polishing wheel. A somewhat tedious task! Needles were not produced from stainless steel and therefore had to be coated with white petroleum jelly to prevent them from staining and rusting.

One of England's first companies to specialize in the manufacture of quality needles was founded by a man called MacKenzie, in Whitechapel, London. MacKenzie's needles soon gained a reputation for being the finest, strongest and best in the trade. Unfortunately MacKenzie got into severe financial difficulties and was forced out of business by his creditors. Around this time, Alcester, in Warwickshire, became the better-known home of needle production, and trade in London declined.

The following description is reproduced from the *International Hair Route Magazine* by kind permission of the author and editor, Derek Copperthwaite.

Accurate Records from 1852 describe the manufacture of needles as follows:

> A coil of steel wire is cut into lengths sufficient to make two needles. These lengths are collected into bundles, and straightened by a special process. The grinder then takes a number of the pieces in his hand, and points them (at each end of a dry grindstone). They are now washed and dried over a fire.
>
> The next process is done by children: about 50 wire lengths are fastened down to a strip of wood and then passed to a workman at another machine, powered by a treadle under the man's foot, where a file is moved over the needles to remove any imperfections.
>
> From here the needles go to a kind of vice, and the upper part of the double-ended needles are worked backwards and forwards until they break in two in the middle. The tops of the heads are now filed round, and any roughness removed.
>
> Placing them in a furnace until they are red-hot then hardens the needles. From there they are emptied into a tub containing oil or water and then tempered by being placed over a slow fire and allowed to cool gradually. Any crooked needles are mixed with oil, soft soap and emery powder, the needles are then wrapped in a sack-cloth and placed in a kind of mangle, worked by mill power, to be scoured. This process takes about a week, and when done the needles are washed in hot water and dried in sawdust. Winnowing and sorting follow. The points are then set and the needles polished on a leather buffing wheel.

At the time this description was written, upwards of 10 000 people in England were employed in the needle-manufacturing industry. Altogether, in its making, each simple needle passed through the hands of 70 or more workmen and underwent ten or more separate operations. Today's needles are produced in factories using automated equipment and thankfully the procedure is considerably less labour-intensive.

Prior to 1981, when the first pre-sterilized needles were launched, electrolysists' main concerns were the durability of the needles; how long they would last; and what was the most effective method of sterilization for re-use. The autoclave that we have today was not available. The most common practice was to turn the electrolysis machine up to full intensity. A piece of cotton wool was soaked in surgical spirit and then held with a pair of tweezers. The needle was then inserted into the cotton wool between the tweezers and a short burst of high-frequency current was applied to it. Needless to say with the advent of AIDS and hepatitis this method is no longer acceptable. The arrival of sterile, disposable needles has proved to be an asset to electrolysists in the UK and many other countries. Local authorities within the UK insist on the use of disposable needles before granting a licence to practise.

The way electrolysists work today was changed in 1981 when Sterex International launched the first pre-sterilized, disposable epilation needle. This was shortly followed by pre-sterilized needles from Arand Ltd with Ballet needles, and the Carlton Professional needle of Taylor Reeson Laboratories.

Features of pre-sterilized needles include:

- Medical-grade packaging.
- Sterilization by gamma irradiation or ethyl oxyene gas.
- Needles packed individually either in single pouches or tear-off strips.

The main reason for the interest in disposable needles was the possible cross-infection from HIV and hepatitis. Manual sterilization of needles is time-consuming, and there is a risk of inadequate sterilization if the procedure is not carried out thoroughly. Up to February 1990 there have been three documented cases of acupuncture-related AIDS. Sterex introduced the first sterile disposable needle in 1981, the needles initially being imported into England from the USA. These needles were then packaged individually and sterilized by gamma irradiation. Owing to the increase in demand, Doug Cartmell and John Heath began the manufacture of Sterex needles in England. These are now exported widely. Needle diameters range from 003 to 006, with the addition of 010 for the treatment of warts and skin tags.

During 1988, the 'Ballet' needle was launched by Joseph Asch of Arand Ltd. Before entering the epilation market, Arand had specialized in the exportation of high-quality sterile needles for use in medical surgery and acupuncture. Many prototypes were designed before Arand were finally satisfied with the quality.

The Ballet needle is constructed of a single piece of highly polished stainless steel, packed in hospital-grade blister packs and sterilized with ethylene-oxide gas. For sensitive skin this needle is also available in 24 carat gold plate, bonded evenly onto the needle surface. The needle diameter is available in 002, 003, 004, 005 and 006. The 002 is very fine, proving particularly useful for facial and upper-lip work or where the client is worried about fine, vellus hair.

In February 1989 Carlton Professional entered the market with a two-piece flexible needle, diamond drawn from austentic stainless steel and finished with a final polishing process. Each needle is fitted with a protective plastic cap, individually packed and sterilized by gamma irradiation. Needle diameters are 003, 004, 005 and 006. This needle took 19 months to develop as opposed to the 6 months originally envisaged.

CTI Medical Equipment introduced a range of disposable needles during the summer of 2000. The needles are available in stainless steel or gold plate. The needles are packed in strips of five and sterilized by gamma irradiation. The range of needles includes an angled needle in sizes 002, 003, 004 and 005 and a flexible two-piece needle, sizes 002, 003, 004 and 005.

For electrical epilation, a good-quality needle should have a smooth, polished surface (see Figure 9.1), to facilitate insertion into the follicle. When the surface is smooth, the current will flow evenly to the tip of the needle. This is to ensure that the current flows to the point of least resistance. Where the surface is rough (see Figure 9.2), the current dissipates at the roughened surface, therefore less concentration of current reaches the tip of the needle where it is required. Current intensity will have to be increased to destroy the papilla and lower follicle effectively. This results in a more painful treatment for the client.

The next consideration is the shape of the needle point/tip. Tests have shown that with a micro-polished, rounded point (see Figure 9.3), current intensity can be lowered significantly. When the needle point is rough, micro-lesions to the follicle can result, which hinders the healing of the skin after treatment.

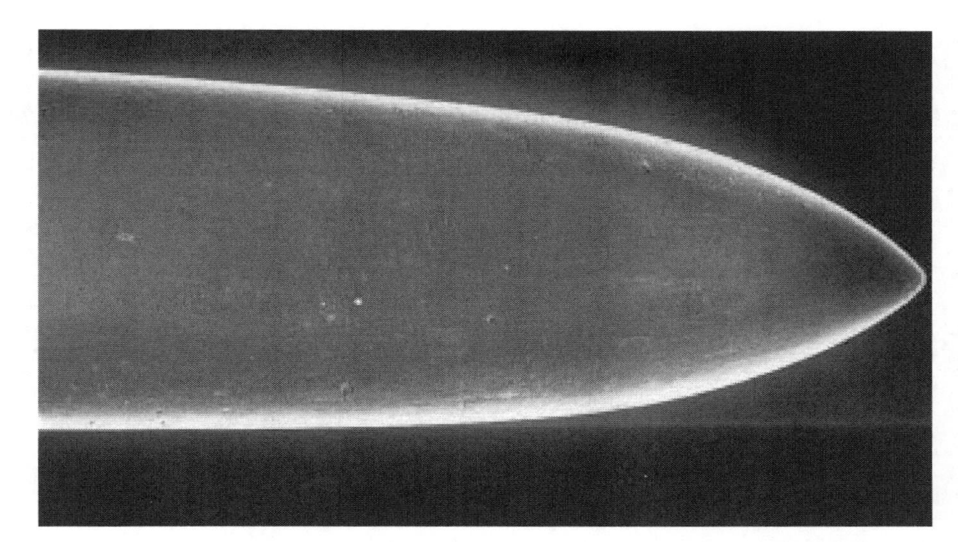

Figure 9.1 Smooth needle surface

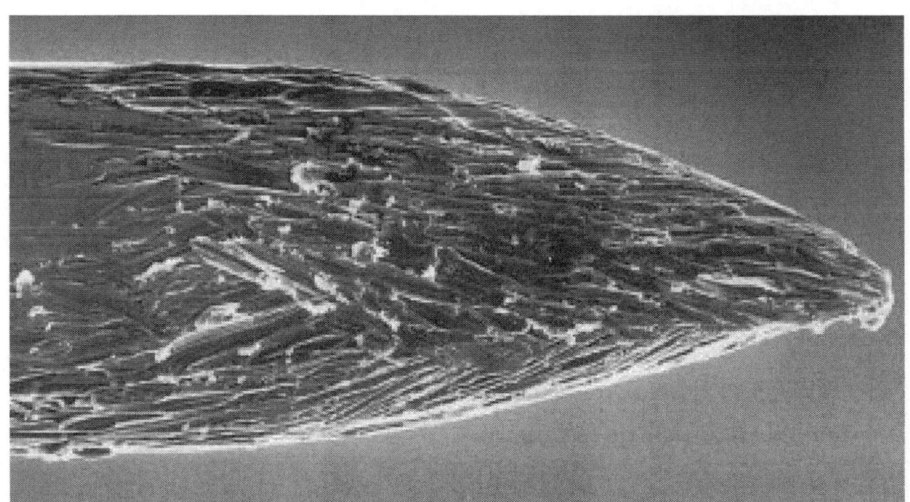

Figure 9.2 Rough needle surface

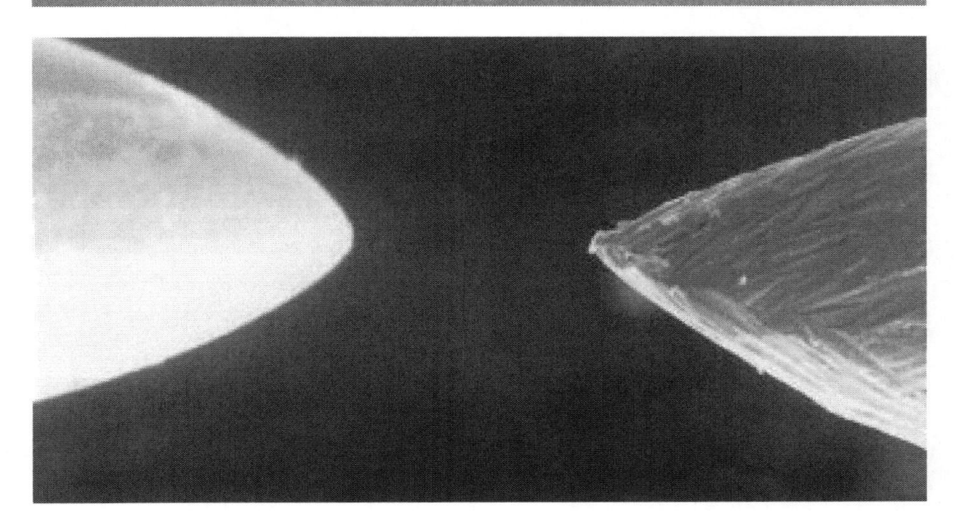

Figure 9.3 (*left*) Rounded point; (*right*) sharp point

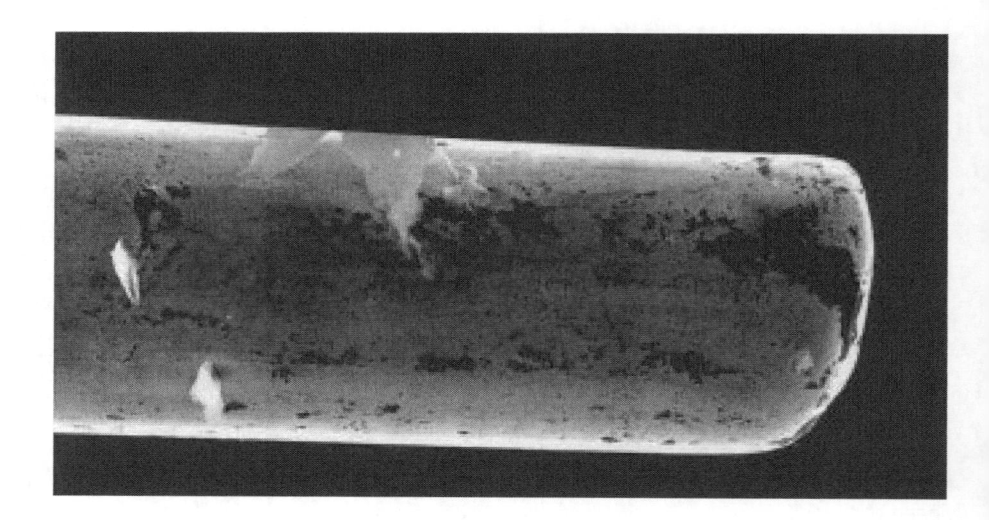

Figure 9.4 Blunt needle

When the point of the needle is blunt (see Figure 9.4), easy entry into the follicle is hindered, whereas when the point is too sharp the risk of piercing the follicle wall or probing through the base is increased.

Further considerations when choosing the needle are the diameter, length and shape. The diameter of the needle should match the diameter of the hair, i.e. the coarser the hair, the larger the needle and conversely the finer the hair, the smaller the needle. This enables the tip of the needle to encompass the base of the follicle, so enabling the current to reach the entire area requiring treatment. When the diameter of the needle is too small, the intensity during high-frequency treatment is concentrated on a small area, which is more painful for the client. There is also a risk of under-treating the follicle, due to insufficient current reaching the entire base. When the needle is too large, the follicle wall could be stretched, resulting in current dissipating at the skin's surface instead of at the base of the follicle. The follicle wall may be stretched, causing broken capillaries or bruises.

Needles can be obtained in both regular and short lengths. It is not always possible to reach the base of deeper follicles when the needle is too short. This means there will not be enough current reaching the target area.

There are eight types of needle available:

1 One-piece tapered.
2 Two-piece – straight.
3 Two-piece – tapered.
4 Insulated.
5 24 carat gold-plated.
6 Uni-probe.
7 Angled.
8 One-piece pointed needle for advanced epilation.

One-piece tapered needle

This needle is constructed from one-piece stainless steel, which is polished to give a smooth surface (see Figure 9.5). There is less risk of the needle breaking or separating from the shank. When the needle is tapered, the current – be it galvanic or high-frequency – is concentrated at the narrowest point, which will be the tip of the needle. In other words, there will be low current density at the larger area and higher current density at the tip – where the action is needed. The risk of over-treating the upper follicle and epidermis is therefore reduced.

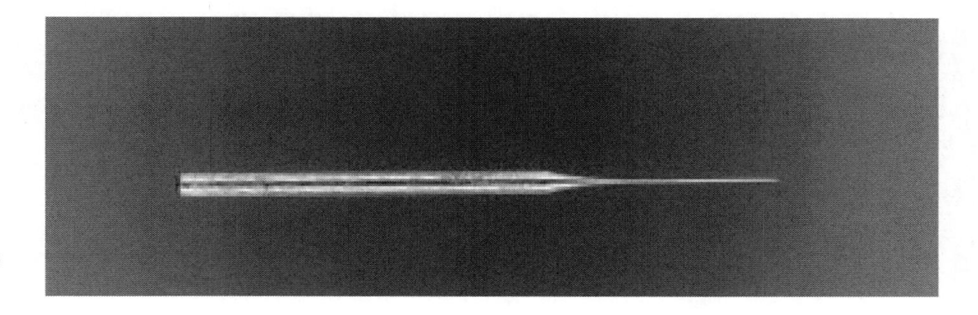

Figure 9.5 One-piece needle

These needles are available in two shank sizes – F and K. The K shank is more widely used in European countries and partly in American and Australian needle holders.

Two-piece needle

The two-piece needle (see Figure 9.6) is constructed from stainless steel using a fine piece of steel running through a stainless steel shank. This construction is more flexible. The two-piece needle is available in both straight and tapered shapes. Availability varies in different countries.

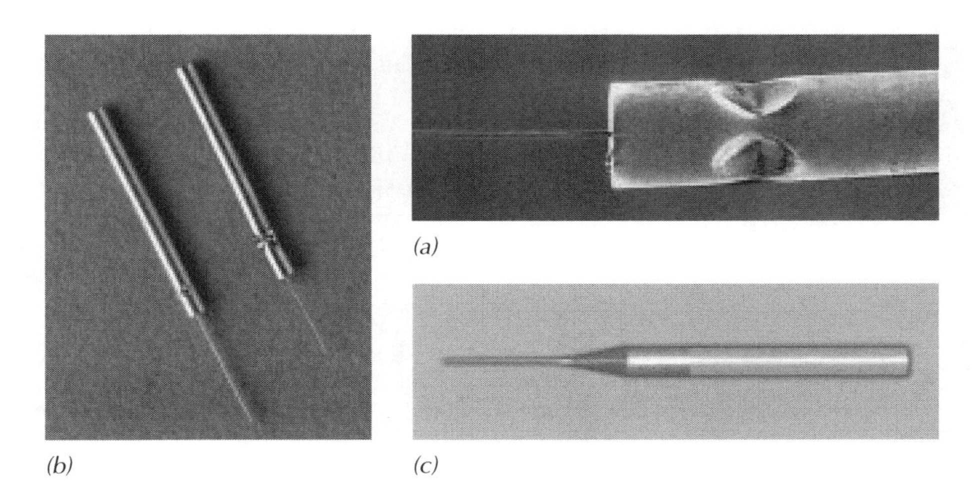

(a) (b) (c)

Figure 9.6 *(a), (b)* Two-piece needles; *(c)* one-piece needle

Insulated needle

The insulated needle consists of a needle coated with an insulating material which leaves 1/25 inch (1.0 mm) of the needle exposed (see Figure 9.7). Traditionally, the disadvantage of insulated needles is that the insulation makes the needle thicker, and the insulation is inclined to lift away from the needle after sterilization, so hindering insertion. There may also be a slight risk of deposits of insulation material in the hair follicle.

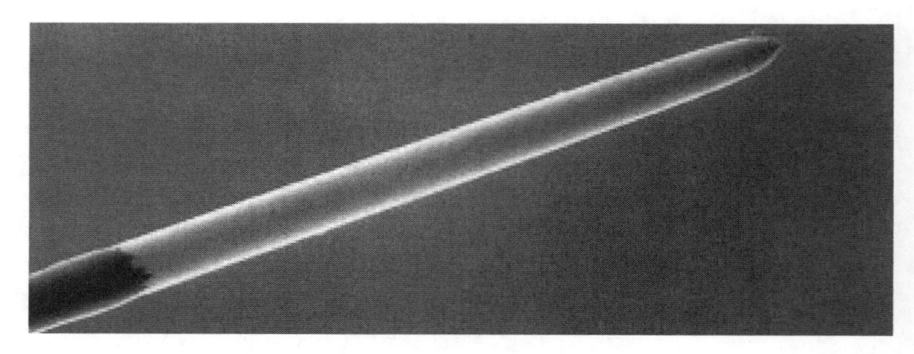

Figure 9.7 Insulated needle

In the spring of 1991 Ballet introduced the first disposable insulated needle. The insulating material is a modern polymer that is only 0.00004 in (1 μm) thick. This minimal addition makes an imperceptible difference to the electrolysist.

The advantage of insulated materials is that the current is concentrated at the tip of the needle, and is therefore ideal for sensitive skin. Because the needle is only used once and the polymer is extremely smooth and hard, the insulation remains on the needle.

Gold-plated needle

The gold-plated needle consists of a stainless steel tapered needle which has had 24 carat, gold plate bonded on to the entire needle surface. The addition of cobalt gives gold a little bit of strength. Gold is an excellent conductor of electricity and does not oxidise with age. It is a slippery metal, which aids ease of insertion.

This needle is ideal for people who are sensitive to stainless steel. Swelling and erythema are reduced – both important benefits in all electrical epilation treatments, including the treatment of telangiectasia.

Uni-probe

The Uni-probe is a combined disposable needle and cap (see Figure 9.8), which does away with the need to wash, rinse, dry and sterilize caps. It also eliminates the risk of accidentally bending the needle when inserting the needle into the cap. To make identification easier, each needle size is manufactured in a specific colour.

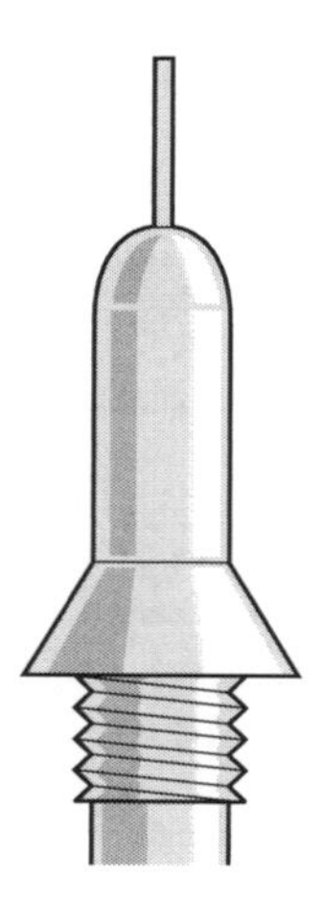

Figure 9.8 Uni-probe (E.A. Ellison)

Angled needle

The angled needle (Figure 9.9) is ideal for treatment of:

- Telangiectasia.
- Removal of skin tags.
- Epilation to the bikini line and under the chin.

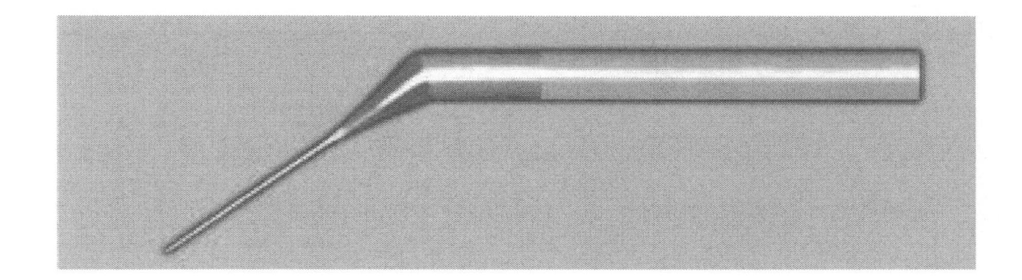

Figure 9.9 CTI's angled needle

Telangiectasia needle

The qualities of a needle used for removal of telangiectasia and minor skin blemishes differ from those of an epilation needle used for hair removal. In contrast to an epilation needle, which ideally possess a slightly rounded point, a telangiectasia needle requires a thin tapered point which will pierce the skin easily without causing trauma during entry into the capillary. Figure 9.10 shows the difference between the point profile of the two needles. The telangiectasia needle is thin and gradually tapered with a sharp point – ideal for easy and gradual penetration to the target depth; the electrolysis needle is more rounded – for smooth insertions into a follicle without risk of penetration.

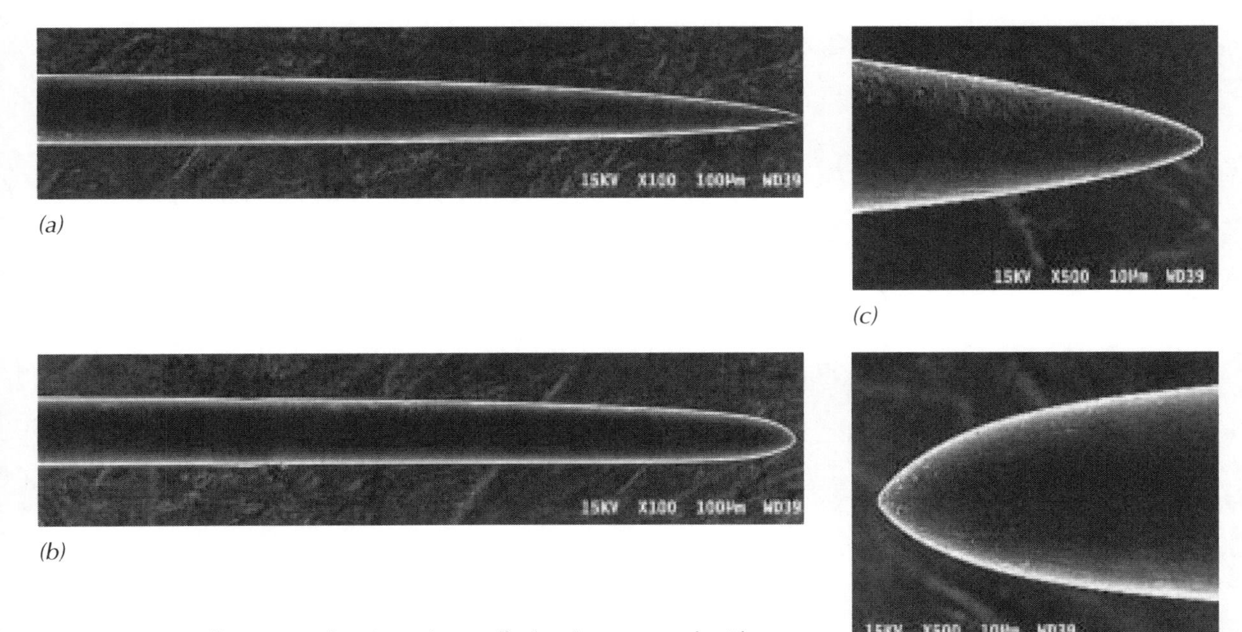

Figure 9.10 Ballet's TEL telangiectasia needle (*a, c*), compared with a Ballet electrolysis needle (*b, d*)

Needles are available in stainless steel and gold finishes. Gold is an excellent conductor therefore the operator may find that current intensity can be reduced.

Review questions

1. Name the two methods used for sterilizing disposable needles.
2. State the advantages of using disposable needles.
3. What are the requirements of needles when used for electrical epilation.
4. What effect does a rough needle have on current distribution?
5. Why should the diameter of the needle match the diameter of the hair?
6. State the purpose of an insulated needle.
7. What are the advantages of a gold-plated needle for sensitive skin?

10 Electricity

Electricity has a number of uses. It may be used to produce heat, cold by refrigeration, light or chemical changes by various processes, e.g. galvanic electrolysis. Without electricity, electro-epilation cannot take place.

The wise electrolysist will be familiar with the basic principles and applications of electricity for a number of reasons:

1. Safety and efficiency may be increased, with less risk of damaging equipment, giving or receiving an electric shock, or overloading the system.
2. By understanding the effects of applying galvanic current and high frequency to the hair follicle and skin, the operator is able to decide which method of electro-epilation will be most suitable for the area and type of hair to be treated.
3. By understanding the process taking place in the follicle and skin during current application, it is possible to avoid damage to clients through over-treating, or incorrect application.
4. Additional benefits may be the ability to carry out minor repairs, such as changing fuses, repairing cables to needle holders, changing plugs etc. The need to call in an electrician for minor repairs may be eliminated, as may costly bills and loss of income through equipment being out of commission. It should be stressed, however, that unless the electrolysist is indeed skilled in these areas a suitably qualified technician or electrician *must* be called in.

What is electricity and how can it be used to advantage in the salon?

Electricity is referred to as either static or dynamic. It is *static* electricity which builds up on our hair after vigorous brushing, so causing the hair to crackle and fly away; it is also static electricity which results in a mild shock from an object, such as a light switch or car door in which a build-up of excess electrons has occurred. This form of electricity is said to be 'at rest'.

Dynamic electricity is a controlled flow of energy formed from the movements of electrons. It is electricity in motion, e.g. direct and alternating currents.

A simple definition by explanation of the electrical terms commonly used with most equipment in galvanic treatment is given below.

An *electrolyte* is an alkaline or acid solution which allows the conduction of an electric current. Moisture in the skin, which is mildly acidic, enables the current to pass from the needle into the skin tissue. Tissue fluids in the body are electrolytes.

An *electrode* is the conductor, which is used to make contact with the electrolyte. During electro-epilation the needle acts as the electrode and allows the current to flow into the moisture within the hair follicle and skin tissue. When using a direct current, two electrodes are necessary to form a complete circuit.

A direct current possesses *polarity* because it flows in one direction only through a complete circuit. To achieve this the current requires two poles (one negative and one positive) to form a circuit. The electrically charged electrons that form the current flow from negative to positive (see Figure 10.1).

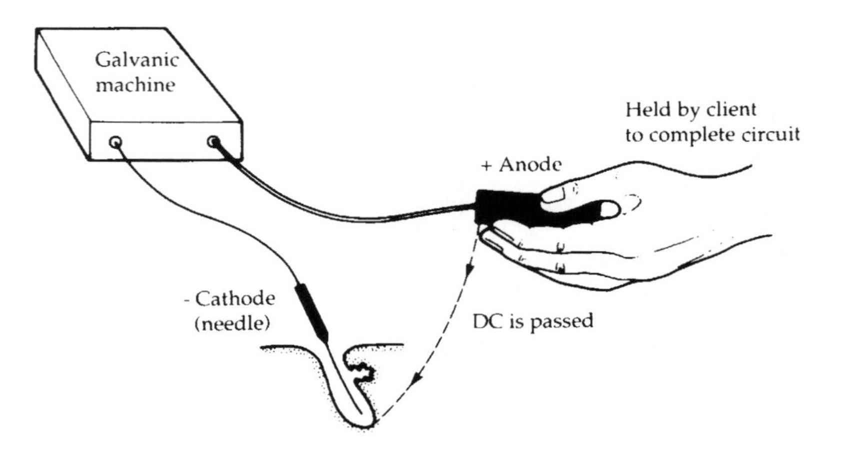

Figure 10.1 Direct current from machine to client

Ionization occurs when a direct current is passed through an electrode into an electrolyte. During electrolysis the salt molecules contained within the body tissues separate into electrically charged ions. The electrolyte solution will allow the movement of positively charged cations towards the cathode (negative electrode) and the movement of negatively charged ions towards the anode (positive electrode).

The *anode* is the positively charged electrode. During electrolysis the client holds the anode. It is known as the indifferent electrode. The anode is connected to the positive outlet of the galvanic machine.

Cataphoresis occurs when the active anode repels positively charged cations into the skin. This procedure may be carried out using either a roller, rod or tweezer electrode. It is usually applied at the end of an epilation treatment by galvanic electrolysis.

The *cathode* is the negatively charged electrode which becomes active during electrolysis. The cathode is connected to the negative outlet of the galvanic machine.

Anaphoresis is the result of the active cathode repelling negatively charged anions into the skin. As with cataphoresis, a roller, rod or tweezer would be used, usually at the beginning of an epilation treatment.

The term *electrolysis*, when used in connection with permanent hair removal, refers to a chemical process which takes place when a direct current is applied to tissue salts and moisture contained within the hair follicle and surrounding skin tissue. The resulting sodium hydroxide produced destroys the tissue because it is highly caustic.

Having followed the current from the machine, through the needle holder and probe into the follicle, consideration should be given to the process that takes place. This makes it necessary for the electrolysist to understand about atoms and molecules etc. These may be defined as follows.

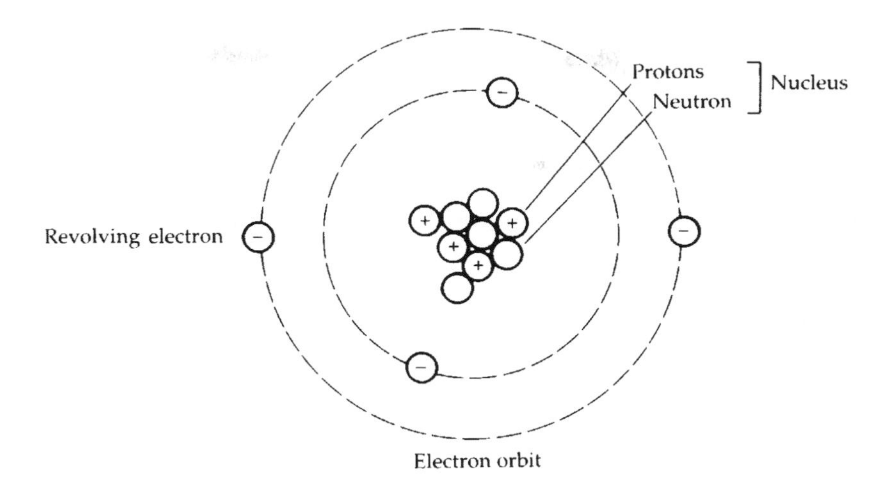

Proton = positive charge
Electron = negative charge
An atom contains an equal number of protons and electrons
An atom is neither positively nor negatively charged

Figure 10.2 The atom

a Sodium chloride (salt) molecule

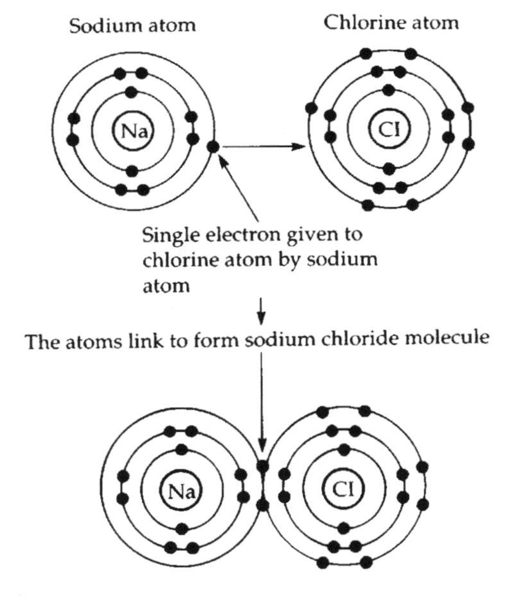

b Water molecule

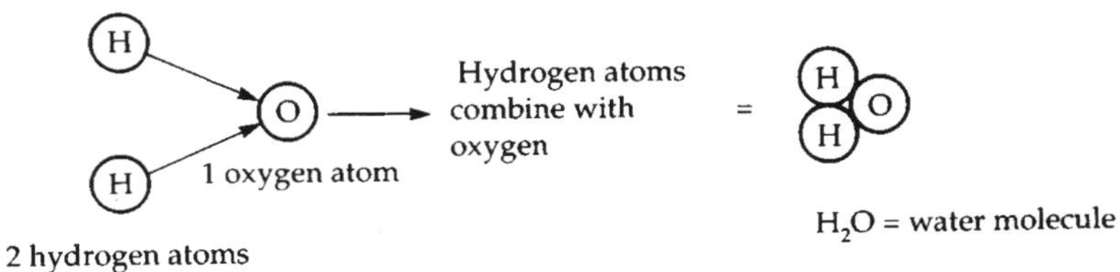

Figure 10.3 The formation of a molecule

Atoms consist of a central nucleus containing neutrons and positively charged protons. Neutrons are electrically neutral. Usually the number of neutrons equals the number of protons. The overall number of neutrons does not affect the electrical charge of the atom. Negatively charged electrons rotate around the protons. When the number of electrons and protons are equal, the atom is neutral (see Figure 10.2).

Molecules are the smallest units into which a substance can be broken down without losing its basic properties. It is possible to divide molecules into atoms. The properties of the individual atoms may be very different to those of the molecule they form, e.g. two atoms of hydrogen and one atom of oxygen combine to become one molecule of water. The properties of water are entirely different from those of hydrogen and oxygen in their separate entities (see Figure 10.3).

Ions are formed as a result of an atom either losing or gaining an electron. A positively charged cation is formed when an atom loses an electron; whereas a negatively charged anion is formed when an electron is gained.

The basic principle of attraction and repulsion can be used to explain the exchange of electrons:

Electrons – negative
Protons – positive

Opposites attract, whereas like repels. Therefore an electron will be attracted to a proton, whereas two electrons will move away from each other (see Figure 10.4).

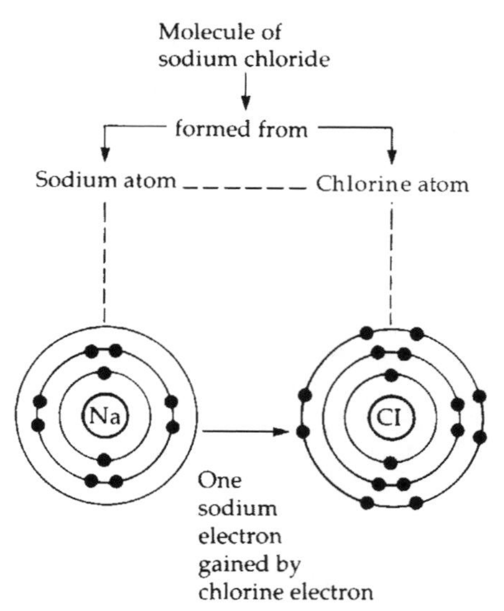

Figure 10.4 The formation of ions

Electric currents used in electro-epilation

Direct current (dc) is a flow of electrons along a conductor in an electric circuit of constant voltage. A direct current possesses polarity. The current flows in one direction and destroys hair by chemical reaction within the tissues.

Alternating current (ac) reverses its direction of flow at regular intervals, thereby changing polarity within the circuit (unlike direct current which flows in one direction only). A *cycle* refers to one complete alteration, or change of direction; the *frequency* means the number of complete cycles per second; and *Hertz* is the term used for a complete cycle (see Figure 10.5).

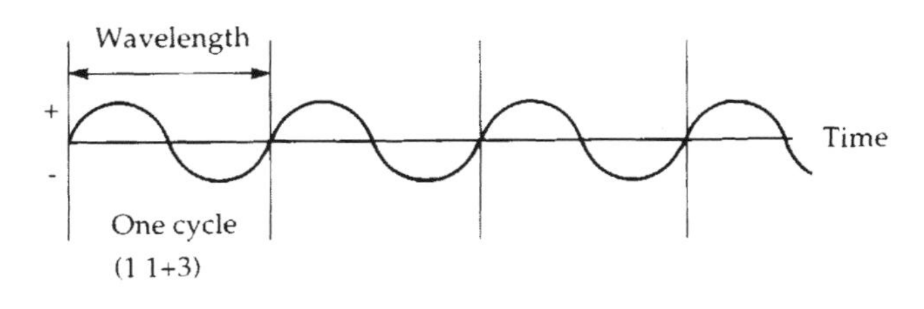

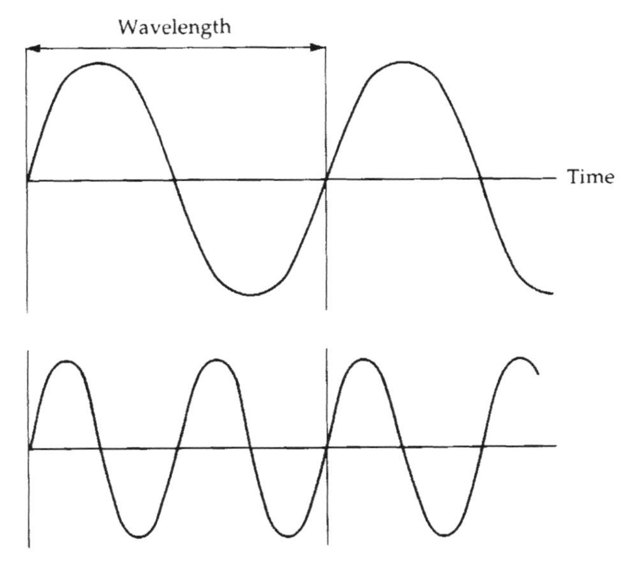

Figure 10.5
Alternating currents

High frequency is used in electro-epilation to destroy tissue by the heat it produces. It is a high-frequency, low-voltage, alternating current, ranging from 3 to 30 million cycles per second, or 3–30 MHz. Theoretically, a higher frequency, lower voltage should produce a more comfortable treatment for the client. In practice much depends on the sophistication of the circuit design.

Components used in electro-epilation equipment

The most central part of equipment is the power supply. This is used to convert electricity from the domestic supply into that of the right voltage and type, i.e. ac or dc, for the rest of the electronic or electrical equipment within the appliance (see Figure 10.6).

The basic components used in a simplified power supply are shown in Figure 10.6. More modern designs will usually have more sophisticated forms of voltage regulation and may use a technique called a switch-mode

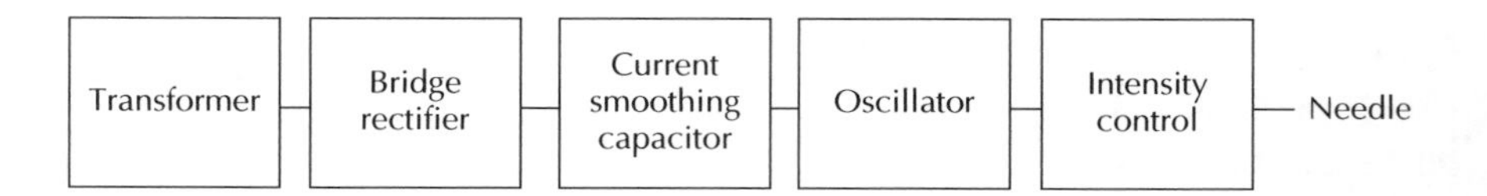

Figure 10.6 Components used in electro-epilation equipment

regulation. However, for the purpose of understanding most equipment and the basic components, this example is adequate.

The *transformer* changes the voltage of an alternating current. It can either step-up/increase the voltage, or step-down/decrease the voltage, and is generally used to convert the domestic supply voltage to that required by the equipment.

The *rectifier* is a device consisting of two or more diodes which changes an alternating current to a direct current. A diode allows the current to flow through it in one direction only. The bridge rectifier, containing four diodes, is more efficient.

The *capacitor*, which may also be known as the filter or the condenser, is used to smooth out the bumps in a direct current that has been produced by a rectifier.

The *voltage regulator* prevents the voltage surging, or suddenly increasing if the client's resistance becomes lower during the treatment.

The *rheostat* (variable resistor), also known as the variable current control, regulates the output of current from the machine to the needle (usually by controlling the voltage regulator circuit).

The *constant current generator* maintains the current at the pre-set level despite any change in resistance (of the client being treated).

Conductors and non-conductors

A *conductor* is a substance such as steel, copper or, in the case of electro-epilation needles, stainless steel or gold, which allows an electric current to flow when an electric pressure is present in a given direction. Other examples of conductors are zinc, carbon and impure water (see Figure 10.7).

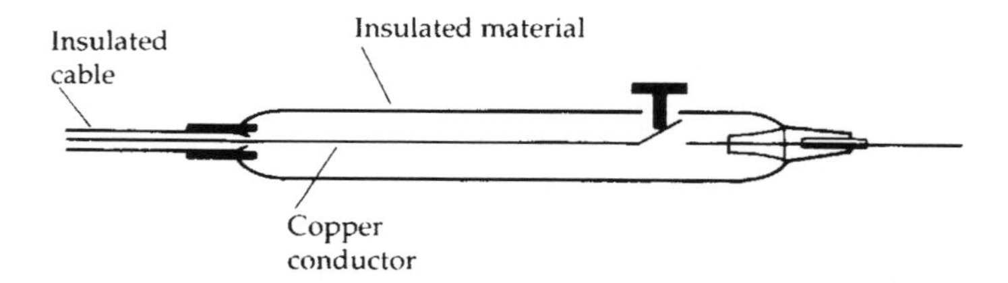

Figure 10.7 Flex and needle holder indicating a conductor (copper wire) and insulator (covering material)

A *non-conductor* prevents the current from passing *along* it; examples are wood, plastic, rubber and glass. A non-conductor can also be referred to as an insulator.

The *cable* of the needle holder demonstrates the use of both a conductor – which is the copper wire contained in the centre – and a non-conductor, or *insulator*, which is the plastic coating surrounding the wire.

Circuits/circuit breakers and fuses

A circuit can be defined as the path taken by an electric current. The circuit may be closed to allow the current to flow or open to prevent it flowing.

Fuses, switches and other overload devices may be used intentionally to open the circuit and prevent the current flowing in the event of a malfunction such as a short or equipment failure, thereby protecting both the client and the operator (see Figure 10.8).

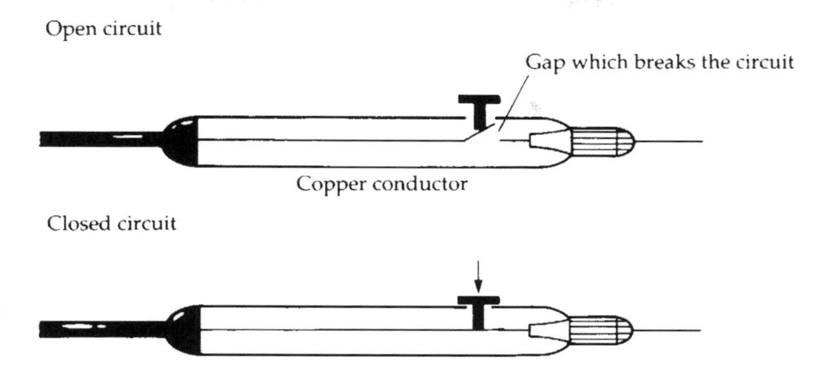

Figure 10.8 Open and closed circuits

An open circuit exists when the flow of current has been interrupted, usually by a switch, but sometimes as a result of broken wires contained within the flex.

Circuit breaker

This is a switch that breaks the flow of current when the circuit becomes overloaded. Its function is to prevent damage to the circuit or equipment through an overload of current. The switch is activated by thermal contact. When the intensity of the current is too high, the circuit is broken. The circuit breaker serves the same function as the fuse, the difference being the switch can be reset whereas the fuse must be replaced.

Fuses

The fuse is designed to be the weakest link in an electrical circuit. When too high an intensity of current is passed, the fuse will blow (melt), thereby breaking the circuit and preventing further current flow. The purpose of the fuse is to act as a safety device to prevent an overload of current to a piece of equipment, or possibly damage to another part of the wiring. A faster-acting and safer method of protection is the earth leakage contact breaker.

The *cartridge fuse* is used in all flat-pin plugs. It consists of a porcelain or glass tube containing the fuse wire soldered onto a metal contact cap at each end. The fuse is held in position by metal clips connected to the live terminal. When the fuse is blown the whole cartridge is replaced. The correct fuse value should be used at all times. The most widely used fuse values/ratings are 3 amps, 5 amps and 13 amps.

The older-type of *mains fuse* box contains porcelain fuse holders. A piece of fuse wire is connected to a screw at each end of the holder. The fuse rating will depend on the type of circuit it is protecting, e.g. lighting, power sockets or electric cooker. When this type of fuse is blown, only the wire is replaced. It is always advisable to keep a card of fuse wire and a screwdriver next to the mains fuse box. Later boxes have thermal/magnet circuit breakers and cartridge fuses, which may be reset manually.

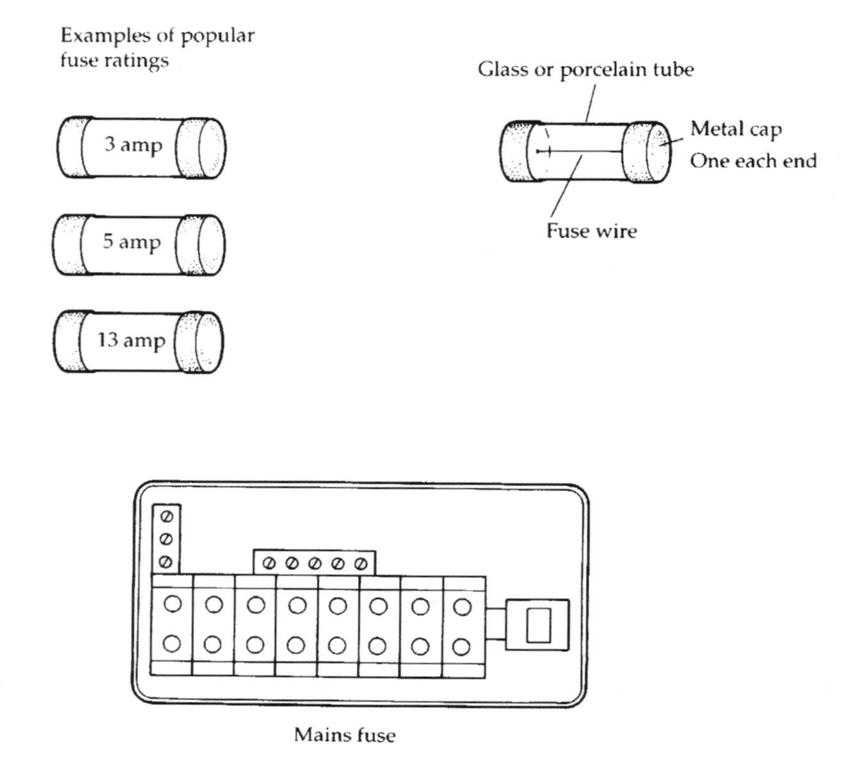

Figure 10.9 Cartridge fuses

Wiring a three-pin flat plug

The neutral wire is blue and should be connected to the left-hand side of the plug.

The live wire is brown and is always connected to the terminal next to the cartridge fuse.

The earth wire is green and yellow. It is connected to the terminal at the top of the plug. The purpose of this wire is to act as a safety device, taking any excess or stray electricity to earth, so preventing any person who is using the equipment from receiving a shock.

Not all appliances contain a three-core flex. Some are designed with 'double insulation', allowing them to function safely with a two-core flex,

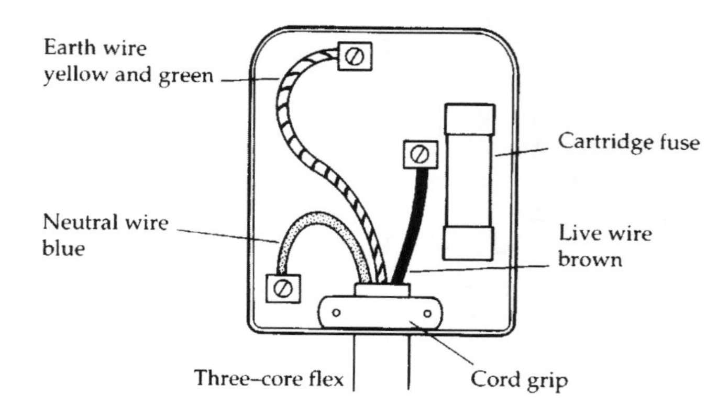

Figure 10.10 Wiring of a three-pin flat plug

which will contain a blue and a brown wire for connection to the neutral and live terminals. *Where an earth wire is provided, it must be connected on a three-wire flex.*

The cord grip on a plug is designed to hold the outer casing of the flex firmly and help prevent the wires from working loose. The wiring in a plug should be checked regularly to ensure that the wires are all still firmly connected to the relevant terminals.

Units of electrical measure

Ampere—Measurement of the intensity of the current. It is the rate of electrical flow and is expressed in coulombs. The strength of the current is equivalent to a flow of one coulomb per second equals one amp.
Coulomb—the measurement of an electrical charge.
Milliampere—One thousandth of an ampere. It is the measurement most frequently referred to in electrical epilation.
Watt—A unit of electrical power. One watt of power exists when one ampere is felt at a pressure of one volt.
Volt—A measurement of electrical pressure.
Voltage—Measures the potential difference between two points of a circuit. The force produced by a battery or generator pushes the electrons, and thereby the current, round the circuit.
Ohm—The measurement of resistance. Conductors offer less resistance to electrical pressure than non-conductors.
Ohm's Law—It takes one volt to push a current of one ampere through a conductor with a resistance of one ohm. This may be expressed as $V = IR$.

Review questions

1 Define the following terms:
   (a) Electrolyte
   (b) Electrode
   (c) Polarity
   (d) Ionization.
2 Describe each of the following:
   (a) Atom
   (b) Molecule
   (c) Ion.
3 What is the difference between ac and dc?
4 Name the components found in an electrical epilation machine and briefly describe the purpose of each part.
5 Define the following:
   (a) Circuit
   (b) Circuit breaker.
6 Compare the cartridge fuse with the mains fuse.
7 Draw and label a diagram to show the wiring of a three-pin plug.
8 What is the function of a cord grip in a plug?
9 Define the following units of electrical measure:
   (a) Ampere
   (b) Milliampere
   (c) Volt
   (d) Voltage.
10 Define Ohm's law.

11 Galvanic electrolysis

The permanent removal of unwanted hair by means of electrical epilation owes its reputation to the thoroughness of galvanic electrolysis. The history of electrolysis can be traced back to 1875 when Dr Charles Michel of St Louis, Missouri, reported to medical colleagues that after applying a negatively charged galvanic current to a hair follicle the hair did not re-grow. In time it was realized that the follicle had been permanently destroyed by the action of sodium hydroxide.

Figure 11.1
Dr Charles Michel

Skin specialist Dr Hardaway followed this discovery in 1876 by using electrolysis for non-medical purposes. The year 1880 saw the publication of *Electricity in Facial Blemishes* – the author Plymouth S. Hayes, MD. Hayes indicated that it was possible to guarantee the permanent removal of hair by means of galvanic electrolysis.

During 1886 Dr George Henry Fox wrote a further publication entitled *The Use of Electricity in the Removal of Superfluous Hair*. Dr Fox was one of the more sympathetic medical practitioners of that time where superfluous hair was concerned. He was aware that although excessive hair growth did not kill the patient, it certainly had a detrimental effect on her mental health and well-being. However, galvanic electrolysis had a number of drawbacks. Application time of current to the follicle took between one and three minutes to remove the hair. This made the treatment slow and tedious as well as impractical for an extensive hair growth.

The machines used in those early days were not as refined as those used today. The direct current was produced by batteries, and the intensity was often higher, therefore the treatment was more painful for the client. A greater degree of skill was needed to operate these machines correctly to avoid skin damage or an excessively strong current being used. An extensive build-up of sodium hydroxide occurred in the tissues owing to the length of time the current needed to flow. The result was widespread tissue destruction, and in many instances scarring.

In 1916 Professor Kree developed the multiple needle technique. This method speeded up the process by using up to 10 needles at any one time. The needles that were used then were much larger, with rougher surfaces.

Owing to the many drawbacks, galvanic electrolysis was gradually replaced with short-wave diathermy.

The galvanic current

The galvanic current, discovered by Luigi Galvani, is a *direct current* (see Figure 11.2). It is a flow of electrons along a conductor in an electric circuit. When a direct current passes through an electrolyte, which contains ions, the ions move in opposite directions. The ions carry the current.

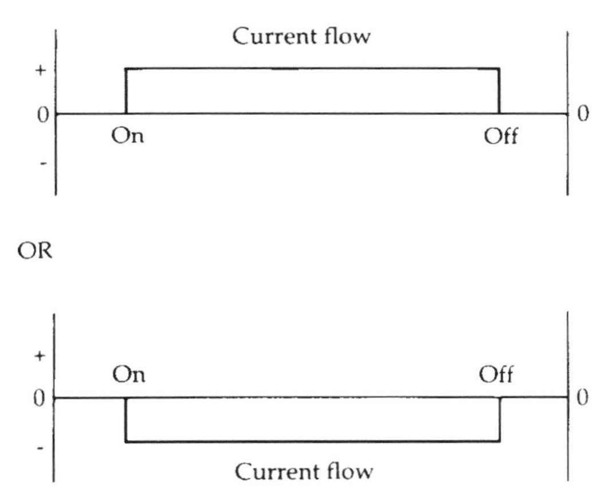

Figure 11.2 Direct current

Electrolysis can be defined as a chemical process that occurs when a direct current is applied to:

1 tissue salts and moisture; or
2 salt-water solution.

The direct current causes the salts and water to split into their chemical elements, which then rearrange themselves to form entirely new substances. During the passage of direct current through salt-water solution the negative chloride ions (anions) are attracted to the positive anode, and the positive sodium ions (cations) are attracted to the negative cathode. When the chloride ions reach the anode they lose an extra electron to become chlorine atoms. Sodium ions arriving at the cathode gain an electron and become sodium atoms. The sodium atoms react with water to form sodium hydroxide, while the chloride ions form hydrochloric acid.

The electrodes used during electrolysis are known as the anode, which is positive, and the cathode, which is negative. Negatively charged anions are attracted to the positive anode, whereas positively-charged cations are attracted to the negative cathode. The principle is that like poles repel one another and unlike (or opposite) poles attract each other.

How does galvanic electrolysis take place?

During the application of galvanic current to the lower follicle chemical changes take place (see Figure 11.3). Moisture and body salts are composed of molecules. These molecules are formed of atoms, which divide under the influence of galvanic current. The ions regroup and are converted into sodium hydroxide, hydrogen gas and chlorine gas.

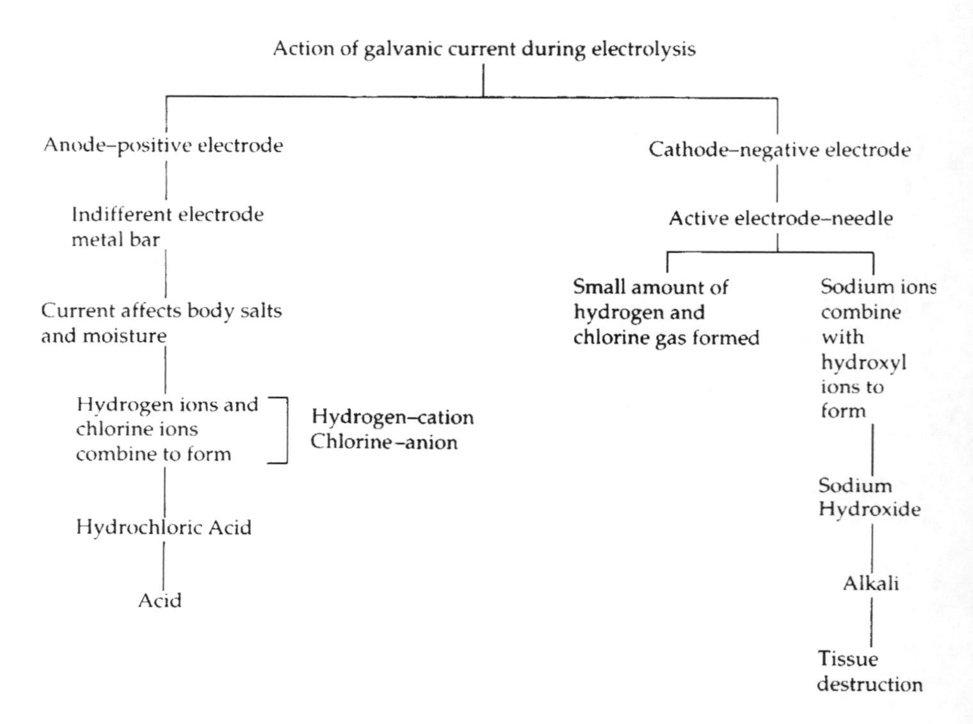

Figure 11.3 Action of galvanic current during electrolysis

This process takes time to develop in the follicle. Sodium hydroxide is not produced immediately but is dependent on the intensity of the current used, together with the length of time the current flows. Therefore:

current intensity × length of application = amount of sodium hydroxide

One-tenth of a milliamp of current flowing for one second will produce one unit of sodium hydroxide. Therefore the longer the time and/or the greater the current intensity, the more sodium hydroxide will be produced. The electrolysist needs to bear this principle in mind when selecting the current intensity and time for application. Full details are given in Chapter 13.

All hairs of a similar type may be treated with one setting. Obviously, more sodium hydroxide is needed for coarse, deep, terminal hairs and less for fine, medium or shallow hairs. Size of hair, moisture content of skin, and stage of hair growth all have an effect on the treatment.

Effect of electrolysis on the hair follicle

When using galvanic electrolysis, destruction of the hair and follicle is achieved in the following way: the indifferent electrode is attached to the positive outlet and handed to the client. This may be covered in damp viscose or damp lint. The needle holder is connected to the negative outlet.

A small amount of negatively charged current is applied to the lower follicle through the needle. When the current encounters moisture within the tissues, a chemical action takes place and sodium hydroxide is formed. This chemical action is not instantaneous, but takes time to develop. It does not stop the moment the needle is removed from the follicle but continues to work for a short time afterwards. The amount of sodium hydroxide produced depends on three factors:

1 The length of time the current is flowing.
2 The intensity of the current used.
3 The moisture content of the skin.

The direct current is available along the entire length of the needle but it affects tissue *only* where moisture is present.

Moisture gradient

As previously mentioned, the presence of moisture is necessary before electrolysis can take place. The skin has the advantage of encouraging electrolysis action to take place at the base of the follicle where it is required due to its natural moisture gradient. The concentration of moisture in the skin is higher in the deeper layers of the dermis, gradually decreasing nearer to the epidermis (see Figure 11.4).

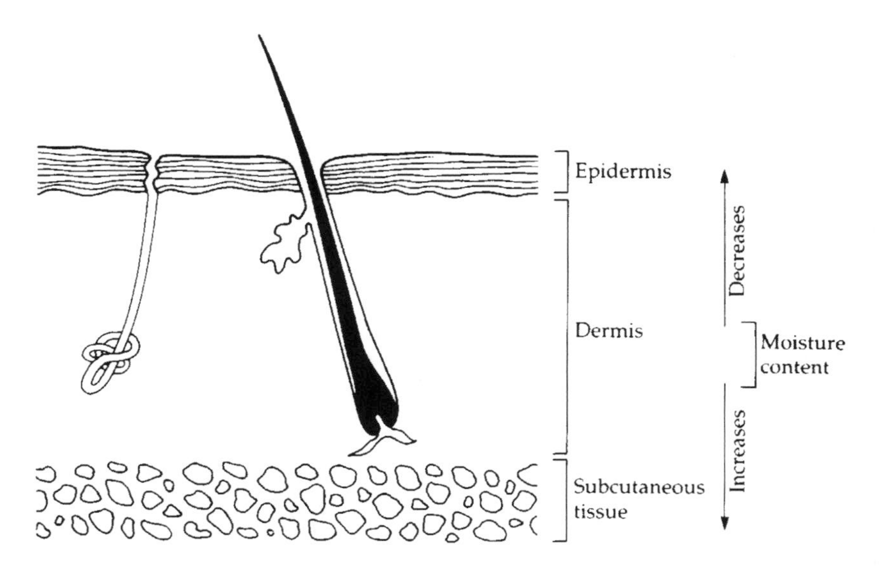

Figure 11.4 Moisture gradient of the skin

Moisture in relation to follicle depth

The hair growth cycle also has a part to play in successful application of galvanic electrolysis. During anagen, the follicle grows down into the deeper layers of the dermis – here the concentration of moisture is good. As the follicle progresses to catagen, the follicle begins to collapse and retreat upwards, thereby moving away from the higher concentration of moisture. In telogen, the active parts of the lower follicle have completely collapsed, leaving only the dermal cord (see Figure 11.5). The resting follicle lies close to the skin's surface, where the moisture content is very low or absent.

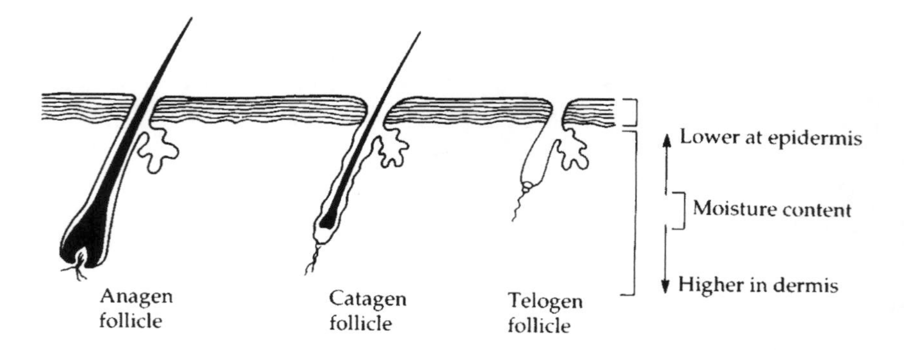

Figure 11.5 Follicles

Vellus hair lies close to the skin's surface. Treatment of this type of hair is more successful with shortwave diathermy rather than blend or galvanic electrolysis, for two reasons:

1 The lack of moisture.
2 The presence of sebum.

Sebum is an excellent insulator. Therefore the sebum present in the upper follicle protects the epidermis from galvanic action taking place close to the skin's surface (see Figure 11.6).

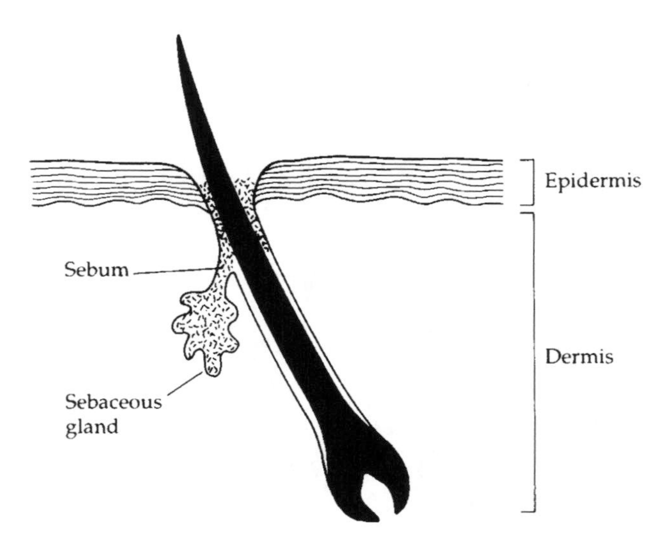

Figure 11.6 Sebum-insulating property

Moisture content and moisture gradient of the skin vary from one person to another and from one area of the body to another. It can be affected by exposure to winds and sunlight, harsh cosmetic preparations, illness or certain medications.

Effects of polarity

The effects of the polarity on the skin will differ depending on how the current is applied.

Current application to follicle using a needle

When using a needle as the active electrode the current is concentrated in a very small area and will result in tissue destruction. It is essential the needle is only ever used with the *negative* electrode (see Figure 11.7).

Needle Negative (-)	**Needle** Positive (+)
Action	
Cathode	Anode
Negative charge	Positive charge
Sodium hydroxide	Hydrochloric acid
Tissue destruction without discoloration	Tissue destruction
	Disintegrates steel needles black oxide deposits
	Tissue discoloration black tattoo marks in skin
Any scar tissue formed will be supple	Any scar tissue formed will be hard

Figure 11.7 Effect of current application to follicle using a needle

Current application to skin surface using rollers

When the current is applied to the skin's surface using a roller or rod (see Figure 11.8), the effects will be less concentrated and give a different result. The cathode (negative roller) will:

1 Soften and relax skin tissues.
2 Irritate nerve endings.
3 Produce erythema due to vasodilatation.
4 Produce skin ionization, causing irritation.

This process, referred to as anaphoresis, can be usefully employed before electrolysis when the follicles are tight. Owing to the relaxing of skin tissues the follicles open slightly, making insertion easier.

When the anode (positive roller) is used, the opposite effects are achieved:

1 Skin tissue is firmed.
2 The superficial blood vessels are constricted, reducing erythema in the area.

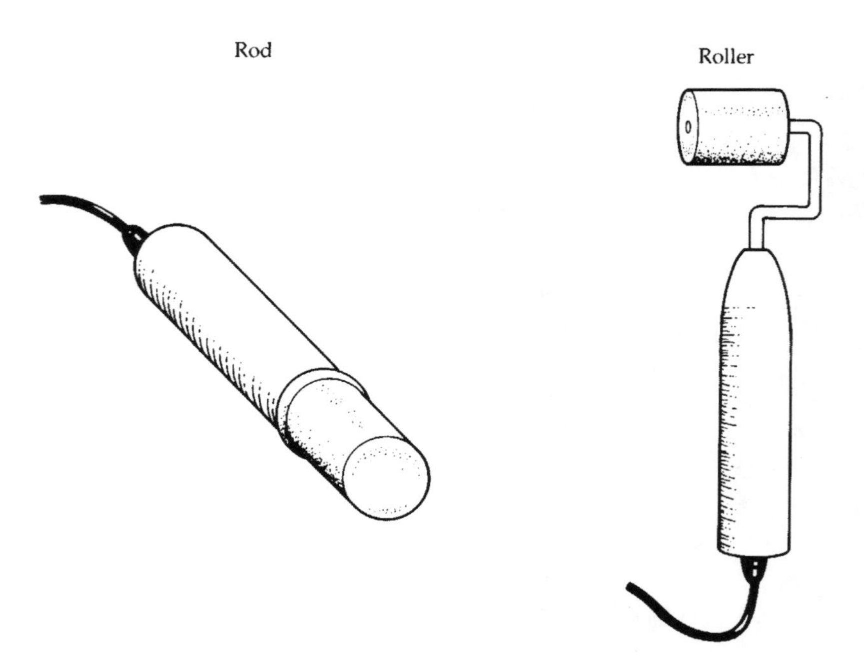

Figure 11.8 Current application to skin surface using rollers or rod

3 Nerve endings are soothed, which is pleasant for the client and induces a feeling of well-being.
4 Hydrochloric acid is formed, which neutralizes the effect of sodium hydroxide and helps to restore the skin's pH balance.
5 Skin de-ionization occurs.

Cataphoresis (using the anode) is beneficial after treatment or when a client's skin is sensitive to galvanic treatment.

Electrolysis equipment

Batteries are no longer used in electrolysis.

1 Transformer.
2 Rectifier.
3 Capacitor.
4 Regulator (voltage).
5 Variable current control.
6 A number of machines contain a meter which indicates the milliamp reading.

Coming from the machine will be:

1 Negative outlet.
2 Positive outlet.
3 Current intensity (milliamp) control.

An alternating current of 240 volts enters the machine and is then changed into a smooth, direct current of up to 100 volts. Most, if not all, electrolysis machines use no more than 30–40 volts dc. This change takes place in the manner described below (see Figure 11.9).

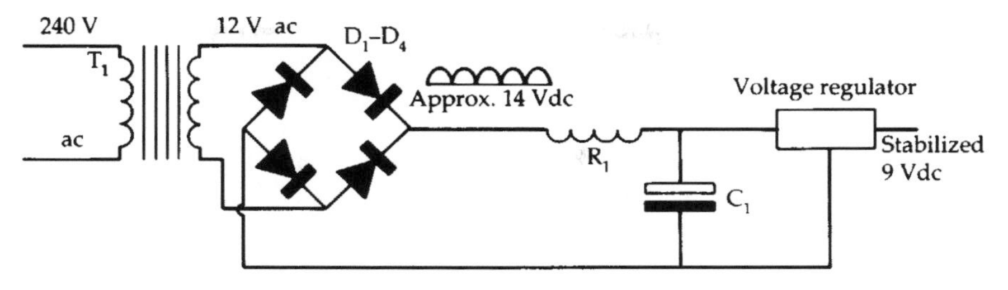

T_1 = Transformer
D_1–D_4 = Rectifying diodes
R_1 = Current limiting resistor
C_1 = Current smoothing capacitor

Voltage regulator or equivalent electronic circuit set to adjust final voltage to that required

Figure 11.9
Alternating current (ac) to direct current (dc)

1. The alternating current is passed through a transformer that reduces, or steps down, the voltage. The transformer also acts as a mains isolation.
2. The bridge rectifier then changes the current from alternating current to direct current.
3. The (filter) capacitor smoothes out any irregularities in the direct current.
4. The filtered/smoothed current goes through a voltage regulator. The constant voltage generator is built in to ensure the voltage stays stable while the current intensity (ma) is varied in accordance with the required setting. The purpose of this is to prevent the current surging or suddenly increasing in intensity should the client's resistance to the current become lower during treatment.
5. The variable current control regulates the current output from the machine to the needle.

Treatment procedure

1. Sanitize the area to be treated.
2. Hand the positive stainless steel electrode to the client, who should be asked to maintain a firm hold to prevent current fluctuation during treatment. The electrode should be covered in damp cotton wool, sponge or lint. This will help the conduction of the current.
3. Insert pre-sterilized needle into the needle-holder.
4. Probe follicle.
5. Apply current, gradually increasing intensity to client's tolerance level. This will normally be between 0.2 and 0.5 milliamps, depending on the type of hair growth to be treated.
6. Check the length of time taken to loosen the hair. This will probably be between 30 and 90 seconds, but may be considerably longer.
7. Remove the treated hair from the follicle.
8. Continue treating hairs of a similar type.
9. Apply aftercare at the end of the treatment.
10. Advise on home care.

When the treatment has been completed, cataphoresis may be applied to the area. For this procedure the negative electrode is handed to the client. The positive electrode, which may be a stainless steel or carbon roller, is moved gently over the skin for several minutes.

The effect of cataphoresis is to reduce erythema (redness); the hydrochloric acid produced neutralizes the effects of sodium hydroxide; and nerve endings are soothed, thereby promoting a sense of well-being and relaxation in the client.

Progressive electrolysis

Progressive electrolysis is the method by which operators work with both hands at the same time (see Figures 13.15–13.17 on pages 110–111). The right-handed operator holds the electrolysis probe in the right hand, the tweezers are held in the left hand. The needle is inserted into the follicle. The galvanic current is activated, the hair is then held with the tweezers (without tension) and the current allowed to flow until the hair releases. The hair is lifted gently every few seconds: if the hair is not ready to release, the tension is relaxed. When sufficient lye has been produced the hair will slide easily out of the follicle. Time, practice and patience are required to master this technique.

Review questions

1. Define 'galvanic current'.
2. Define the term 'electrolysis'.
3. Describe what happens when a direct current is passed through a saline solution.
4. Define the following: (a) anode; (b) cathode; (c) cation; (d) anion.
5. Describe how the sodium hydroxide is formed in the follicle.
6. What is the effect of sodium hydroxide on the follicle?
7. How does galvanic electrolysis take place?
8. Name three factors that affect the production of sodium hydroxide in the hair follicle during galvanic electrolysis.
9. What is meant by the term 'moisture gradient'?
10. How is galvanic electrolysis affected by the moisture gradient in the skin?
11. In what way does sebum affect the application of galvanic electrolysis?
12. Compare the effects of galvanic current applied through the needle on (a) the negative charge and (b) the positive charge.
13. List the effects of (a) anaphoresis and (b) cataphoresis in galvanic electrolysis.
14. What are the benefits of applying cataphoresis after galvanic electrolysis?
15. List the disadvantages of galvanic electrolysis.
16. Describe the treatment procedure for galvanic electrolysis.
17. State the main advantages of galvanic electrolysis.

12 High-frequency treatment

History

The use of high-frequency for electrical epilation came into being because galvanic electrolysis was slow, painful and resulted in a high percentage of scarring. Valuable contributions from many people were responsible for the development of high-frequency and its eventual use in electrical epilation. The principal pioneers included the following:

Heinrich Hertz was the first scientist to demonstrate the existence of high frequency. Hertzian waves were named in his honour.

Professor Arsene d'Arsenoval of Paris, discovered that it was possible to introduce hertzian waves into the body at oscillation frequencies of 100 000 cycles per second without causing muscle stimulation.

Guglielmo Marconi discovered radio waves in 1896 through the radiation of electrical energy into space.

Van Zyneck observed in 1899 that organic tissue could be heated by high-frequency.

Dr Bordier of Paris wrote the first article on the use of high-frequency for the removal of hair.

During the 1940s short-wave diathermy treatment superseded galvanic electrolysis. Initially, short-wave diathermy appeared to be a much faster and more effective method. However, the main disadvantage is the higher percentage of re-growth.

High-frequency voltage

High-frequency is an oscillating alternating current of very high frequency and low voltage. This ranges from 3 to 30 MHz or 3 to 30 million cycles per second. The high-frequency value describes the number of times the current completes a cycle every second. Each frequency has a fixed wavelength. As the frequency increases the wavelength decreases. Thus the wavelength of a 13.56 MHz machine is twice that of a 27.12 MHz (see Figure 12.1).

Owing to the possibility of radio waves generated by electrical epilation units interfering with radio and television transmissions, countries have allocated and standardized specific permitted frequencies. Many countries use 13.56 MHz. CTI's Blendette 48 ACM uses 40 MHz.

High frequency current

+

0

0

-

One
cycle

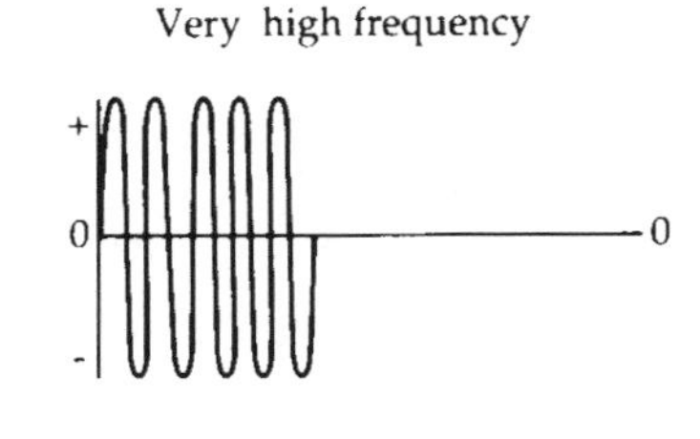

Figure 12.1
Frequencies

Epilation equipment

The following terms are often referred to in electrical epilation:

1 Short-wave diathermy.
2 Radio frequency or RF.
3 Thermolysis, high-frequency or HF.

The explanation for these terms is straightforward. Epilation equipment utilizes the short-wave band radio frequencies. When applied to the follicle, thermolysis takes place that is, tissue is destroyed by the heating effect caused by the agitation of molecules in the surrounding tissue.

An epilation unit consists of the following components (see Figure 12.2):

1 An oscillator, which may be formed from a capacitor and inductor, and/or a crystal.
2 A power supply to drive the oscillator.
3 A variable intensity control.

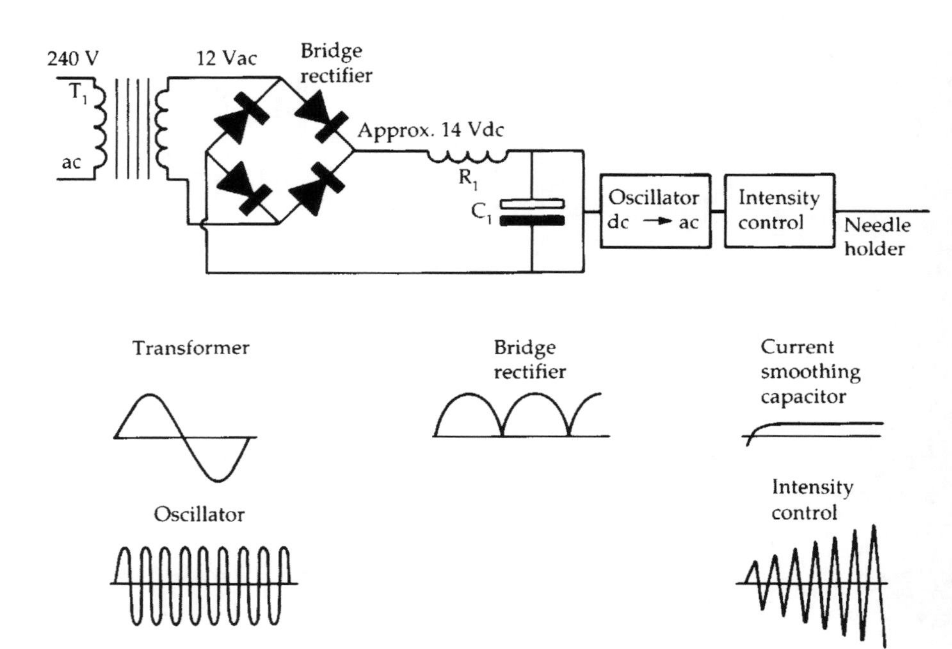

Figure 12.2 Progress of current in high-frequency epilation machine

Crystal oscillators are used in a number of instances, stabilizing the capacitor/inductance design. The advantage of this type of oscillator is that each crystal oscillates at its natural frequency whereas capacitor/inductance type oscillators can be affected by several factors such as humidity, temperature, material and component quality. This means that capacitor/inductance oscillators can have a wide fluctuation in frequency. The tolerance of the inductance/capacitor oscillators also determines the tolerance of the frequency. In contrast, crystal oscillators can have accuracy in excess of 0.001%.

In a capacitor/inductance type oscillator a 'variable' or 'tuning' capacitor is used to vary the coupling of the oscillator to the needle holder output socket in order to control the output intensity of the machine. In a crystal control oscillator the same is achieved by varying the amplitude of the oscillator signal.

Inductance/capacitor type oscillators have a maximum frequency of 18 MHz. In crystal oscillators, where the manufacturer selects a higher frequency, e.g. in excess of 27.12 MHz, factors include the length of the cable to the needle holder, and cable impedance. These factors are critical as they affect the efficiency of transfer of energy from the generator to the needle tip.

Production of heat by high-frequency

During the application of high-frequency to the follicle the electrovalency of molecules within the tissues is altered. The rapid agitation of atoms causes the atoms to vibrate against each other, which results in friction. This in turn causes a temporary release of energy in the form of heat. It is the moisture within the tissues that is heated – not the needle.

The heating pattern commences at the sharpest point of the needle (which should be the tip), where the high-frequency energy is most intense, gradually building up around the needle. The term 'high-frequency field' is used to describe the heating pattern radiating from an epilation needle, connected by a wire to a high-frequency oscillator. The 'high-frequency field' is strongest close to the needle, and in practice will concentrate around the needle tip.

Effect of heat on tissue

Heat destroys tissue either by cauterization or coagulation.

Cauterization occurs when a high intensity of high-frequency is passed into the tissue. The moisture vaporizes and the tissue becomes dry. *Coagulation* occurs when a lower intensity of high-frequency is used. The cellular structure in the tissue breaks down and protein is congealed. Electrical epilation should aim at coagulation of the lower follicle to bring about destruction without damaging the surrounding tissue.

Heating pattern

The heating pattern is the shape of the heated area around the needle during the application of high-frequency. The heating pattern will develop where moisture is present. The ideal development of heat is pear-shaped, commencing at the tip of the needle and gradually building up around it (Figure 12.3).

A number of factors influence the successful application of high-frequency in the follicle:

1. Needle diameter.
2. Quality of the needle surface.
3. Needle depth.
4. Application time of high-frequency.
5. High-frequency intensity.
6. Moisture content and gradient.
7. Correct and accurate insertion.

Needle diameter

The needle diameter should equal the diameter of the hair to enable the tip of the needle to encompass the base of the follicle. When the needle is too small, the intensity will be concentrated on a smaller area. This is often

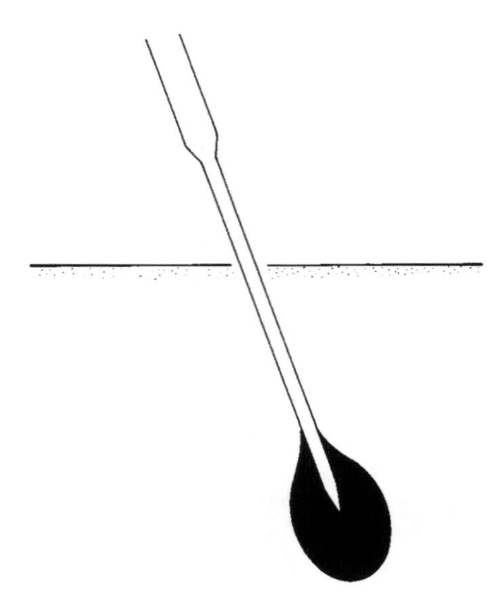

Figure 12.3 High-frequency heating pattern

more painful to the client (owing to concentration of high-frequency to a smaller area), but does not give enough high-frequency to the area being treated. If the diameter is too large, the follicle wall could be stretched which may result in bruising of the skin and broken capillaries (see Chapter 9).

Needle depth

When the insertion is too shallow, the high-frequency is applied too close to the skin's surface. This could result in surface burns and blistering. With shallow insertions, the high-frequency misses the base of the follicle and therefore permanent destruction will not take place. When the insertion is too deep, the base of the follicle will be penetrated and the high-frequency applied to the tissue below, so resulting in destruction of the deeper tissue. This will lead to the development of pit marks and scars (see Figure 12.4).

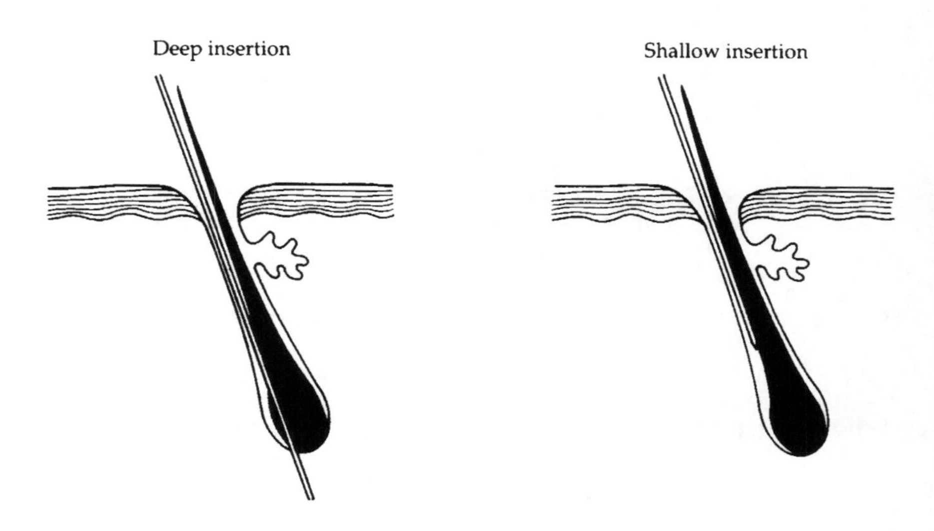

Figure 12.4 Needle depth

Application time of high-frequency

The effect of time on the heating pattern should be considered. The heating pattern starts at the tip of the needle, progresses up the shaft and at the same time expands in width around the tip where the concentration of high-frequency will be highest. When the high-frequency application time is too short, insufficient heat will be generated. When the application time is too long, the heating pattern will eventually reach the surface of the skin, resulting in scars (Figure 12.5.)

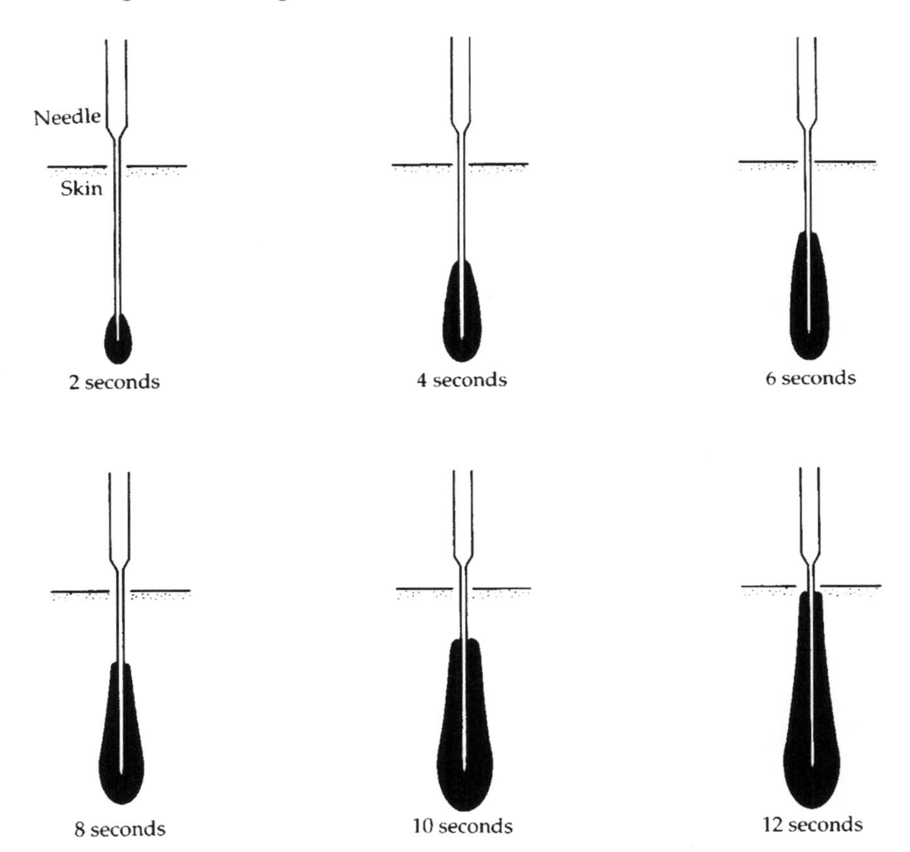

Figure 12.5 The effect of time on high-frequency heating pattern

High-frequency intensity

When intensity is too low, tissue coagulation will take longer, therefore application time will need to be increased. When the current intensity is higher, the heating pattern will commence at the tip of the needle, gradually building upwards. The aim in the application of high-frequency is to set the intensity as high as the client can comfortably tolerate, yet at the same time achieve tissue destruction by coagulation, not cauterisation.

Moisture content of the skin

As with galvanic electrolysis, the moisture gradient has an effect on the application of high-frequency to the follicle. The concentration of moisture is higher in the deeper layers of the skin, so helping to limit the action of the high-frequency to the lower third of the follicle. Problems can occur when the client's skin contains high moisture content at the surface. Skill by the operator is required to keep the electrical action in the lower follicle. With a skin that has high moisture content it is possible for the high-frequency action to reach the surface of the follicle before sufficient destruction has taken place in the lower follicle.

Treatment procedure

1. Sanitize the area to be treated.
2. Insert sterile disposable needle into needle holder.
3. Turn on machine!
4. Probe follicle.
5. Apply sufficient high-frequency to remove hair easily without causing an adverse skin reaction. The current intensity should be set within the client's pain threshold.
6. Remove treated hair with sterile tweezers. The hair should slide out of the follicle without traction.
7. Continue treating hairs of similar type, adjusting high-frequency intensity when necessary.
8. At the conclusion of the treatment apply suitable aftercare.

Flash technique

Flash refers to the application of a very high intensity of high frequency to the follicle for a fraction of a second. Due to the short duration of high-frequency application, the client's nerve endings do not have time to respond, therefore less pain is experienced. The two methods of high-frequency application differ in tissue destruction and heating pattern.

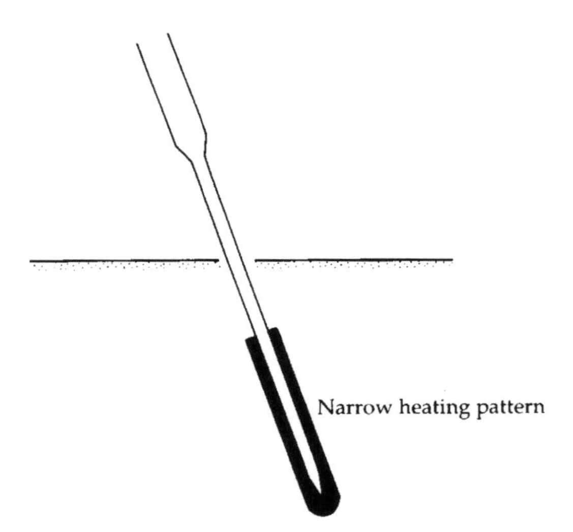

Figure 12.6 Flash technique

With flash technique the heating pattern is narrow, rising up the follicle quickly (Figure 12.6), whereas a lower high-frequency setting produces a pear-shaped heating pattern expanding from the needle tip. Tissue destruction by flash is brought about by desiccation, in other words, the heat produced dries out the moisture and burns tissue, whereas a lower high-frequency intensity applied for a longer period of time breaks down cellular protein and results in tissue destruction by coagulation.

Review questions

1 Define 'high-frequency'.
2 Explain the meaning of the following terms associated with electro-epilation:
   (a) shortwave diathermy
   (b) radio frequency
   (c) thermolysis.
3 (a) Name the main components used in a shortwave diathermy machine and (b) describe the function of each of these components.
4 Explain how the high-frequency current creates heat within the follicle.
5 What is the effect of heat on the skin tissues?
6 Explain how tissue destruction is achieved by:
   (a) coagulation
   (b) cauterization.
7 Draw a diagram to illustrate the ideal heating pattern when high-frequency is applied to the follicle.
8 List the factors that influence the successful application of high-frequency in the follicle.
9 List the disadvantages of using the incorrect needle size during electro-epilation.
10 How does the quality of the needle surface affect the efficiency of electro-epilation treatment?
11 Describe the effects on the skin when the needle depth during treatment is:
   (a) too shallow
   (b) too deep.
12 How does intensity affect application time of the high-frequency?
13 How does the skin's moisture content affect application of electro-epilation?
14 What is meant by the 'flash technique'?
15 What are the disadvantages of using the flash technique?

13 The blend technique

The blend epilation technique is an exciting development that enables the electrolysist to treat superfluous hair more effectively. The high percentage of regrowth experienced with short-wave diathermy is reduced; the length of time taken to clear an area is reduced; and subsequent scarring, which often results when using galvanic alone, is eliminated.

The blend technique came into being as a result of the foresight and efforts of Henrie St Pierre and Arthur Hinkle. Pioneer electrolysist Henrie St Pierre was not satisfied with the length of time taken to destroy hair follicles permanently with galvanic, or the high percentage of regrowth experienced with high-frequency. He joined forces with Arthur Hinkle (author of *Electrolysis, Thermolysis and the Blend*), who at this time was an electronics engineer. Both felt that there must be a way of combining both methods in order to achieve the advantages of each technique.

Figure 13.1 (*left*) Henri St Pierre

Figure 13.2 (*right*) Arthur Hinkle

Work on the blend technique commenced in 1938. A patent was applied for in 1945 and granted in 1948. The technique has now been established for many years in countries such as the USA, Canada, Holland and New Zealand. Britain was slow to follow, with blend being introduced in the early 1980s. However, it was not until 1988 that general interest began to take hold and British companies began the development of blend machines for the UK market.

The principle of blend is to enhance the chemical action of galvanic electrolysis with the simultaneous application of high-frequency.

How does blend work?

The application of the two currents works in the following way: both currents are present at the needle – either separately or together – with each retaining its own characteristics and identity. The dual or blended action takes place within the follicle. The galvanic current brings about tissue destruction by chemical action, whereas high frequency produces heat to speed up the chemical action brought about by the galvanic current. (High-frequency when used in short-wave diathermy destroys tissue by heat.)

The blend is a versatile epilation method that can be adapted to treat most hairs effectively, including curved or distorted follicles and deep bulbous hairs. Blend requires approximately 75% less time to provide the same degree of destruction as galvanic electrolysis alone. It is faster than galvanic electrolysis but slower than short-wave diathermy.

Curved or distorted follicles are unsuitable for treatment by high-frequency (see Figure 13.3). As the tip of the needle cannot be placed accurately at

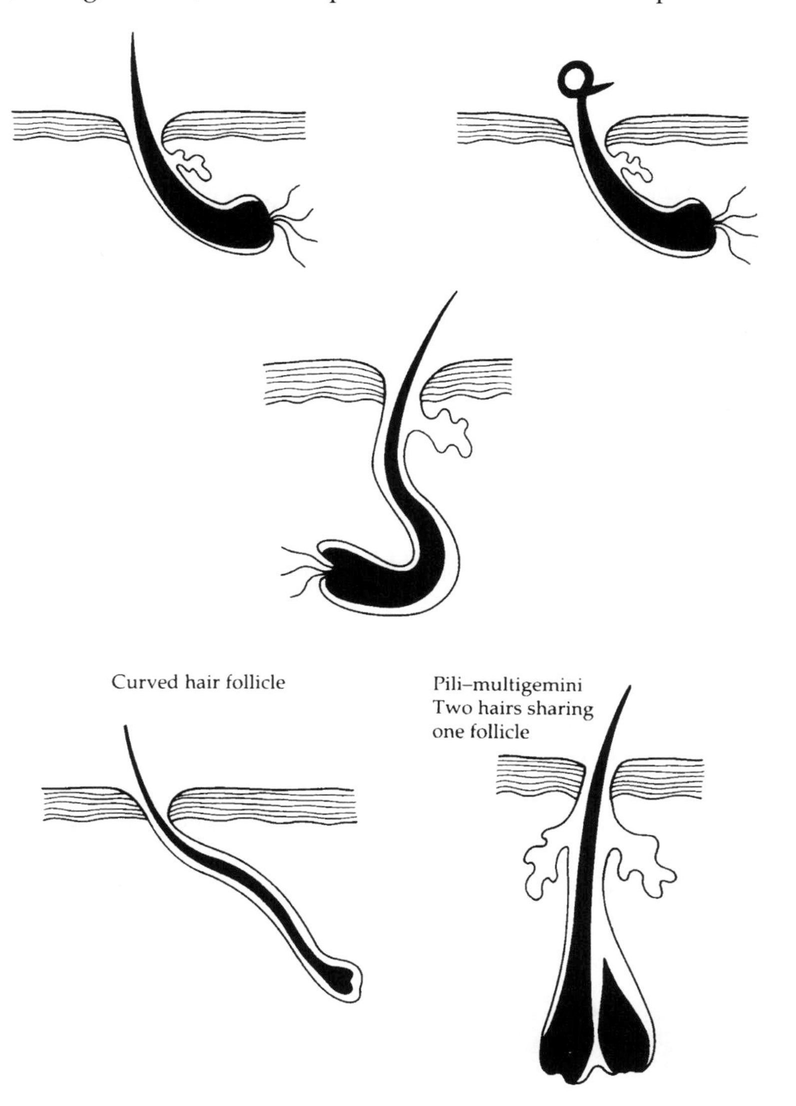

Figure 13.3 Examples of distorted follicles

the base of the follicle, the high-frequency heating pattern cannot reach the base of the dermal papilla and base of the lower follicle. The sodium hydroxide produced during blend, being fluid, can easily move into any open space that will accommodate a liquid.

The successful application of the blend technique relies on the correct balance between the two currents. When used together, galvanic and high-frequency currents are superimposed. Both currents are available at the needle either separately or together as required. Each current *retains* its own true form and is not changed by the presence of the other current. The tissue is affected by each current in the following way:

- *Galvanic current:* electrons flow along the needle into the follicle. When making contact with tissue and moisture within the follicle a chemical reaction takes place. Sodium hydroxide is produced and, being highly caustic, destroys the surrounding tissue.
- *High-frequency current:* the rapid oscillations of the high-frequency current cause vibrations of water molecules within the tissue. This results in friction, which in turn causes heat. *The action of sodium hydroxide is accelerated by heat.*

The action of the two currents is blended in the tissue. The combination of the action from both currents is more effective than the 'sum' of the separate actions.

Action of high frequency during blend application

The action of high-frequency in the current in the follicle enhances that of galvanic electrolysis in three ways:

1. Heat is produced by high-frequency which breaks down the cellular structure of the follicle, causing the protein to congeal. The congealed tissue becomes porous, allowing the passage of sodium hydroxide into the tissue.
2. High-frequency (oscillation) creates turbulence, so forcing galvanically produced sodium hydroxide into the porous tissue.
3. The heat produced increases the action of sodium hydroxide, thereby shortening the time needed to destroy the lower follicle. Heated sodium hydroxide is 2–16 times more effective in destroying tissue than galvanic electrolysis.

When the balance of the two currents is not correct during treatment, the results will not be as good. When too much galvanic current is used, the advantages of high-frequency speed are lost: there is less causticity, porosity and turbulence owing to the decreased heat produced. Over-treatment of skin will result in weeping follicles. Should too much high-frequency be used the effect would be to cauterize the lower follicle, less galvanic action would be produced, and therefore less sodium hydroxide. Over-treating will result in blanched skin.

The combination of the two currents can reduce the application time of galvanic to between 6 and 15 seconds, with the average being 10–12 seconds on most machines. With a number of modern machines the time is reduced to between 6 and 8 seconds. A minimum of 6 seconds is necessary to achieve the blended action of the two currents in the follicle. Less than 6 seconds does not allow sufficient production of sodium hydroxide.

Much depends on the client's pain tolerance. The treatment appears to be less painful than either current used individually. The combined currents seem to have a numbing effect on the nerve endings.

When applying the current, it is advisable to cover the indifferent electrode – which the client holds – with damp cotton wool or sponge. This prevents irritation to the skin caused by the build up of hydrochloric acid that forms at the anode. The thinner the electrode, the more intense the irritation.

The blend epilator

The blend epilator (see Figure 13.4) consists of the following components: for the galvanic part there will be the transformer, rectifier, capacitor, regulator, milliamp meter, negative outlet, positive outlet and current intensity control. The high-frequency will require an oscillator consisting of capacitor or crystal, plus a current-intensity control.

Accessories will be the needle holder and possibly a foot switch; a cable connecting the needle holder to the machine; a metal or carbonised rubber electrode for the client to hold; and a roller for the application of phoresis at the beginning and end of each treatment. Another useful device is an air pump, which, when activated, injects sterile air through the needle holder in co-ordination with the application of the high frequency and galvanic currents. This has a mild sensitizing or numbing effect on the skin.

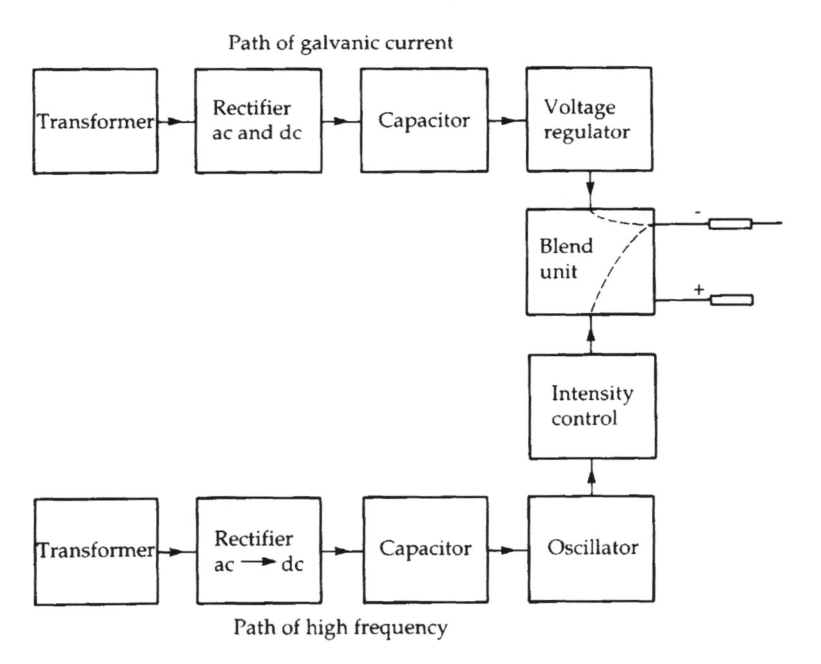

Figure 13.4 Progress of ac and dc during blend treatment

Some machines use two foot pedals to regulate the current, whereas others use one. Others utilize a finger-switch control.

Advanced features found in a number of machines include: automated current settings and timings; insertion sensor; moisture-level test; desensitizing air; galvanic after-count; hair counter; high-frequency timer to eliminate inaccurate counting; and body technique facility.

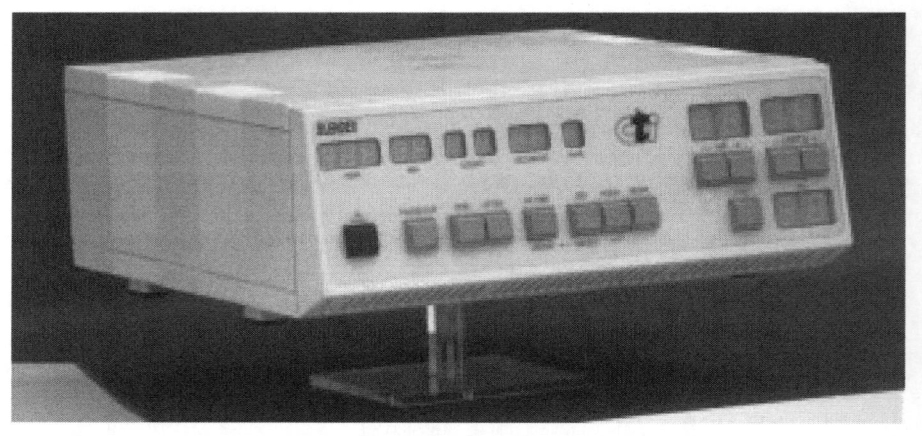

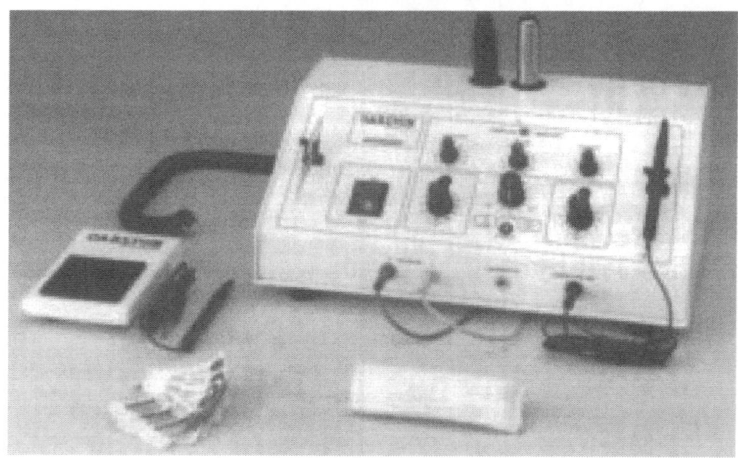

Figure 13.5 *(top)* CTI Blendex; *(bottom)* Carlton Professional Blendette

Units of lye

To make life easier for the electrolysist, Arthur Hinkle introduced the term 'units of lye'. (Lye is the popular name given to sodium hydroxide.) The amount of lye required to destroy a hair follicle permanently will differ according to the type of hair, depth of follicle, stage of hair growth and area being treated. The stronger, deeper hairs require more lye than shallow, vellus hairs. All hairs of a similar type may be treated identically.

The rate of lye production is directly related to the amount of current being used. Lye is not produced instantaneously but takes time to develop in the follicle. A simple formulation, adapting Faraday's law, is:

Tenths of a milliamp × time in seconds = units of lye

i.e. five-tenths of a milliamp × 3 seconds = 15 units of lye.

Electrolysis produces lye on a very small scale within the follicle. It takes one-tenth of a milliamp flowing for one second to produce 'one unit of lye'. Therefore the amount of lye produced in the hair follicle is the product of current intensity used, times the length of time the current flows. The longer the time and/or the greater the current intensity, the more lye will be produced (see Table 13.1).

Table 13.1 Guide to units of lye

Fine, unpigmented vellus hair	10–15 units
Fine, pigmented, soft hair	10–20 units
Medium/shallow terminal hair	15–30 units
Deep terminal hair	30–45 units
Very deep terminal hair	45–60 units

Determining the balance of the blend

The aim of blend epilation is to achieve complete destruction of the lower third of the follicle, including the dermal papilla, in the shortest possible time.

It is the high-frequency current which determines the length of time the blend is applied to the follicle. Care must be taken to avoid using too much high-frequency at too high a setting, which would be to the detriment of the blend technique. The application time is affected by:

1. The client's pain tolerance.
2. The sensitivity of the skin.
3. The type of hair growth, e.g. fine, medium, coarse.
4. The stage of follicle development.

The milliamp setting and the depth of insertion affect the production of lye. A shallow insertion reduces the amount of current flow and therefore the amount of chemical action. With deeper insertions more chemical action will occur owing to the higher concentration of moisture in the lower layers of the skin.

Establishing the current setting

Method A

Determine the number of seconds required to loosen hair with high-frequency. *Using high-frequency only, establish a 'working point' that will give the correct high-frequency setting. This will be a lower setting than used for diathermy.*

Count the number of seconds needed to loosen the hair. Divide high-frequency settings into units of lye to give the galvanic setting in milliamp tenths, e.g. three seconds of high frequency (hf) divided into 15 units of lye (ul) equals five milliamp tenths (ma), or 0.5 milliamps (see Table 13.2).

Table 13.2 Relationship between high-frequency and galvanic settings

Seconds of hf	*divided into*	*Units of lye*	*Milliamp tenths DC*	*Milliamps*
5	divided into	15	3 tenths	0.3
15	divided into	30	2 tenths	0.2
6	divided into	30	5 tenths	0.5
5	divided into	30	6 tenths	0.6
9	divided into	45	5 tenths	0.5
5	divided into	45	9 tenths	0.9
10	divided into	60	6 tenths	0.6
6	divided into	60	10 tenths	1.0

Note: from the client's point of view, with stubborn hairs, it is better to increase the length of time during which high-frequency is applied rather than to increase the galvanic milliamp setting.

Method B

Find the dc threshold level. Insert the probe into the follicle. Turn on dc, gradually increasing the current. When the client's tolerance level has been reached, stop the current flowing and check the milliamp reading. Calculate the length of time needed to treat the hair. Divide units of lye by milliamp tenths, e.g. 0.5 ma on legs requiring 60 units of lye: 60 ul divided by 5 milliamp tenths = 12 seconds.

Variations of blend application

1 Having determined the high-frequency intensity, apply both currents at the same time (simultaneously), and then finish with an after-count of galvanic. The purpose of this is to encourage sodium hydroxide to fill the follicle on removal of the hair. This method is the most widely used in Holland, Germany, the USA, Canada and Japan.

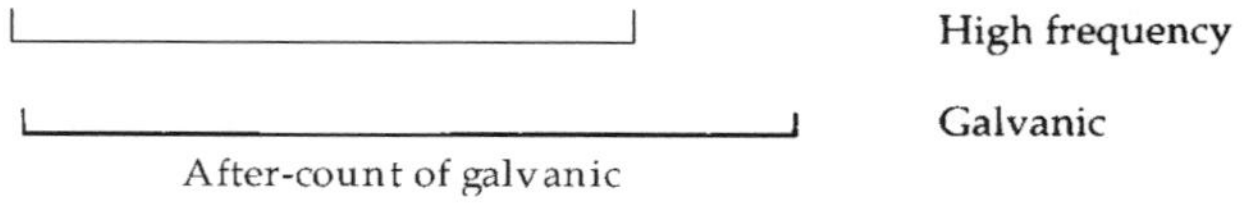

High frequency

Galvanic

After-count of galvanic

Figure 13.6 After-count of galvanic

2 *Simultaneous application of both currents* during the entire application to the follicle.

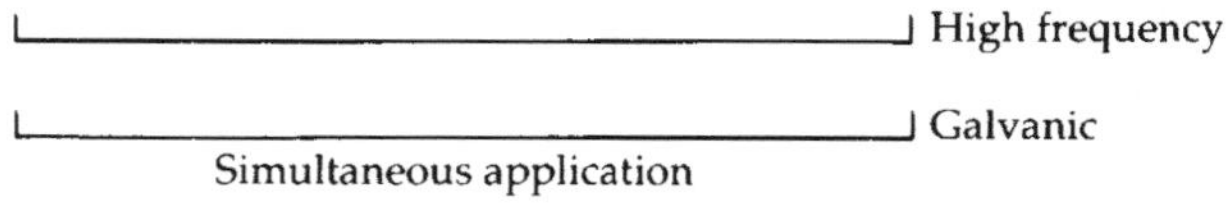

High frequency

Galvanic

Simultaneous application

Figure 13.7 Simultaneous application

3 *Start with high-frequency* then apply both currents simultaneously. The purpose of this is to warm the follicle slightly and also numb the nerve endings. When using this variation it is essential that the high-frequency intensity is not too high, otherwise the moisture in the follicle will be removed and this in turn will prevent the formation of sodium hydroxide.

High frequency

Galvanic

Start with high frequency, follow with simultaneous application

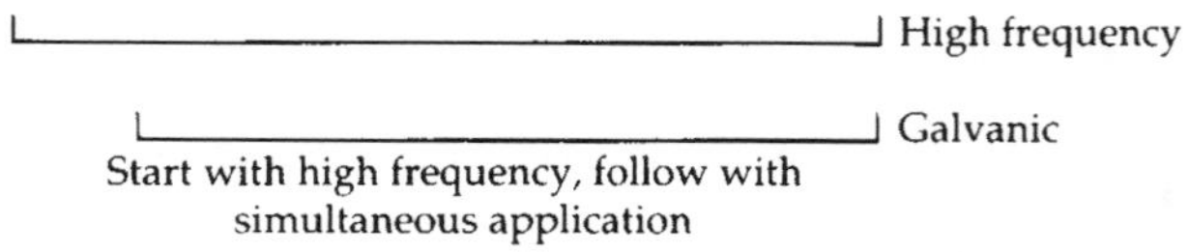

Figure 13.8 Start with high frequency, follow with simultaneous application

4 *Lead with high frequency* to warm the follicle slightly and numb the nerve endings, follow with galvanic current to commence quick production of sodium hydroxide in the pre-warmed environment, and then apply both currents simultaneously.

High frequency

Galvanic

High frequency, galvanic, followed by simultaneous application

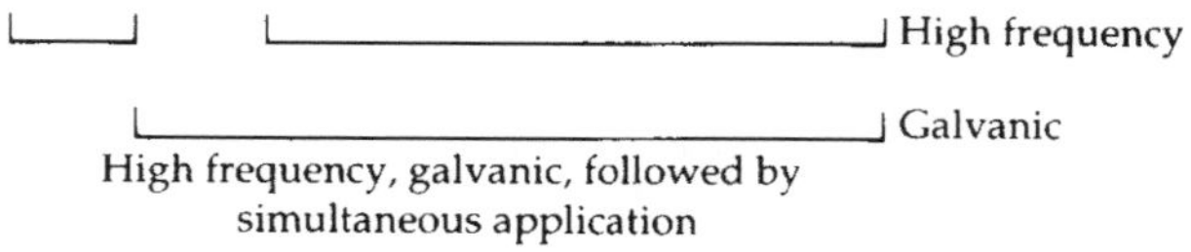

Figure 13.9 High-frequency, galvanic, followed by simultaneous application

5 *Start with galvanic current* to commence production of sodium hydroxide, then follow with both currents.

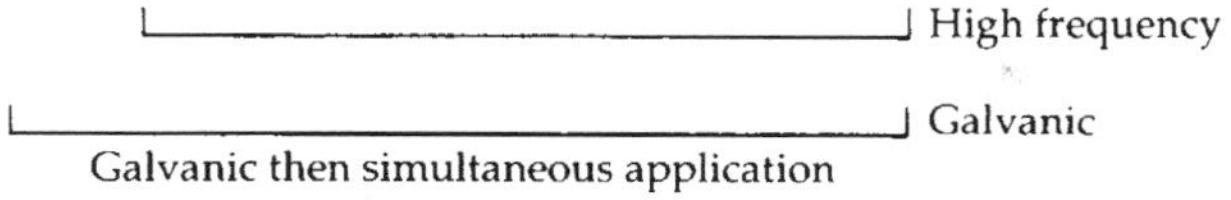

High frequency

Galvanic

Galvanic then simultaneous application

Figure 13.10 Galvanic then simultaneous application

6 *Lead and finish application with galvanic*, apply both currents together in the middle of current application. This last method is a combination of steps (1) and (5).

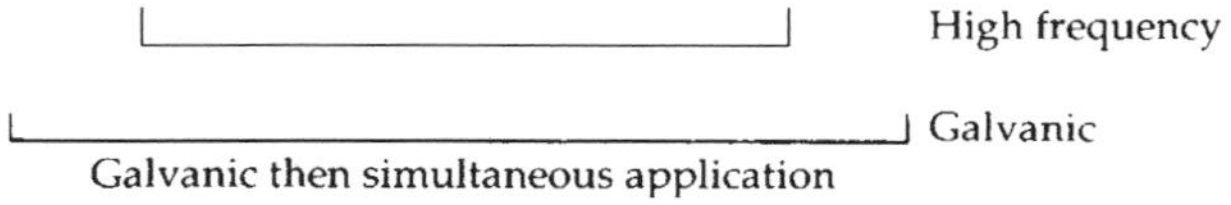

High frequency

Galvanic

Galvanic then simultaneous application

Figure 13.11 Lead and finish with galvanic, with simultaneous application in the middle

Treatment procedure

1. The client should be asked to remove rings and jewellery prior to blend treatment.
2. Remove any make-up or lipstick in the immediate area. Prepare and sanitize area to be treated.
3. Give the sponge-covered positive electrode to the client to hold firmly. The grip on the electrode should remain constant to prevent fluctuating current intensity to the follicle.
4. Examine the skin and select the correct needle size, that is, match the diameter of the needle to the diameter of the hair.
5. Insert pre-sterilized needle into needle holder.
6. Apply anaphoresis if required.
7. Probe follicle.
8. Establish the depth of insertion.
9. Determine the working point. Once the high-frequency setting and application time have been assessed, decide on the units of lye required for the type of hair being treated, to determine the galvanic milliamp setting. (See Tables 13.1 and 13.2.) Treat test hair in chosen manner to determine blend setting.
10. When settings and balance of blend are established, treat all hairs of a similar type.
11. At the end of treatment cataphoresis may be applied. This has a number of advantages: use of the positive electrode has a germicidal effect by producing hydrochloric acid and restoring the skin's pH balance; nerve endings are soothed, which gives a feeling of well-being to the client; erythema is reduced.
12. At the conclusion of the treatment, apply suitable aftercare.

Check that the client has remembered to take all her jewellery with her before leaving the treatment room.

Body technique

The body technique allows the operator to work at a much faster rate by increasing the intensity of high frequency. To allow for the higher intensities, the time must be adjusted according to the client's pain threshold. The galvanic current flows constantly throughout the entire application to the follicle. However, the high frequency is pulsed at one-second intervals, which reduces sensation.

It must be stressed that the body technique is *not suitable for facial work.* This is because the currents are set at a much higher intensity. Honey-coloured crusts appear on the surface of the skin after treatment, which may last up to three weeks. This is due to the increased chemical action that occurs as a result of the higher current settings. When the crusts fall off there is often a faint white mark left behind. While this may be acceptable on the body, it is far from acceptable on the face.

Current application and intensities during treatment

1. Follow the same procedure as that used for treatments to the face and neck.
2. Set the high-frequency intensity according to the client's pain tolerance.
3. Apply the high-frequency current for one second. Test to see if the hair will release. Repeat the procedure. If the hair does not release easily after the two applications, a further one second of current application may be given. Remove treated hairs manually if there is still traction after the third insertion.
4. Once the high-frequency setting and application time have been assessed, decide on the units of lye required for the type of hair being treated, in order to determine the milliamp setting, e.g.

   3 seconds h/f divided into 45 u/l = 1.5 ma galvanic
   3 seconds h/f divided into 60 u/l = 2.0 ma galvanic
   4 seconds h/f divided into 60 u/l = 1.5 ma galvanic

There are a number of variations for current application when using body technique. With time and experience it will be possible to determine the most suitable method of application for each client.

In Figure 13.12 the galvanic current is applied to the follicle without interruption. The high-frequency is pulsed at intervals of one second on and one second off. This has the effect of reducing discomfort to the client while the electrolysist can work at a high speed. The technique is not recommended for the face. The application time and resting interval can be varied according to the requirements of the individual. Machines offering the body technique facility are a great help when applying the currents in this manner.

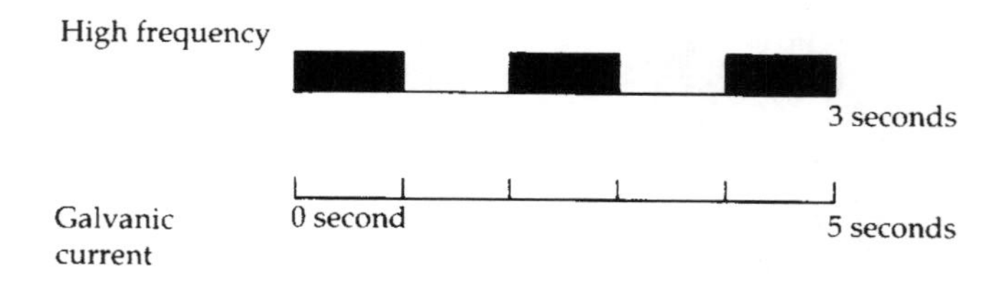

Figure 13.12 Current application during standard body technique

Method B (Figure 13.14) leads and finishes with the application of galvanic current only to enhance lye production.

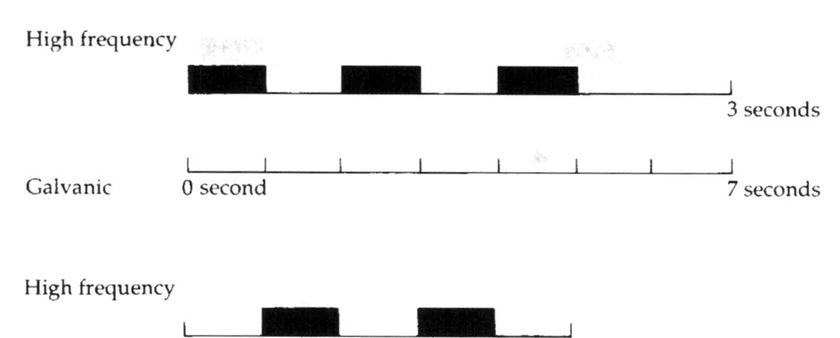

Figure 13.13 Current application, method A

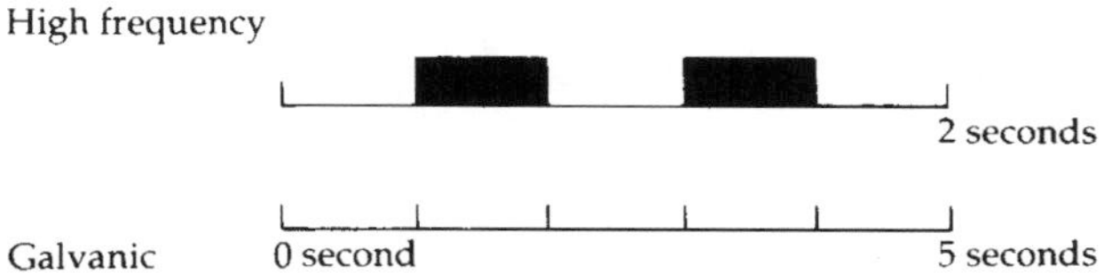

High frequency

2 seconds

Galvanic

0 second

5 seconds

Figure 13.14 Current application, method B

Progressive technique

The two-handed technique, also known as progressive epilation, is the most effective way of destroying the follicle resulting in the minimum of regrowth. This is the preferred method in America, Canada, Holland and New Zealand. For many electrolysists who have been trained in the one-handed techniques used for short-wave diathermy the transition to the two-handed technique is far from easy. However it is worth persevering, for the results are far superior when using the blend technique.

Method of application – progressive epilation

Prepare the client's skin for treatment in the same manner as that used for traditional British blend technique.

Once the balance of the currents has been determined continue in the following manner:

1. The client should be asked to remove her jewellery.
2. Remove any make-up or lipstick in the immediate area. Prepare and sanitize the area to be treated.
3. Give the sponge-covered positive electrode to the client to hold firmly.
4. Examine the skin and hair and select the correct needle size, that is, match the diameter of the needle to the diameter of the hair.
5. Insert a pre-sterilized needle into the needle holder.
6. Establish depth of insertion.
7. Determine working point, determining the balance of the blend and establishing the current setting for the blend.
8. Insert probe into the follicle with one hand.
9. Gently hold the hair with the tweezers with the other hand.
10. Every few seconds apply slight tension to the hair with the tweezers to see if the hair will slide out of the follicle without resistance.
11. Apply galvanic after count to the follicle after the hair has been removed.
12. At the conclusion of the treatment, apply suitable aftercare.

Electrolysists must take care not to touch the needle with the tweezers during the application of high-frequency current. This action would result in a short circuit of the high-frequency and prevent the current from entering the follicle.

Mistakes, which often occur when first attempting progressive electrolysis, are:

1 Taking hold of the hair with the tweezers prior to needle insertion. This action may result in lifting the skin's surface and changing the direction of the follicle, so hindering accurate and correct insertion.
2 Touching the needle with the tweezers.
3 Applying continuous tension to the hair through current application. This action could result in the needle being inserted too deeply into the follicle. This will hinder the release of the hair from the follicle and will give insufficient treatment to the target area.

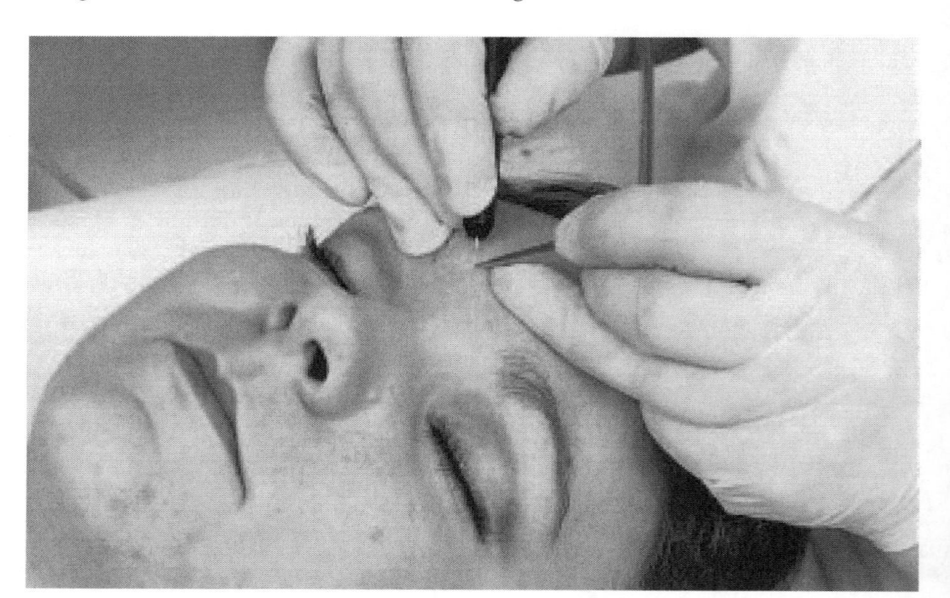

Figure 13.15 Centre forehead, two-handed method, right-handed operator

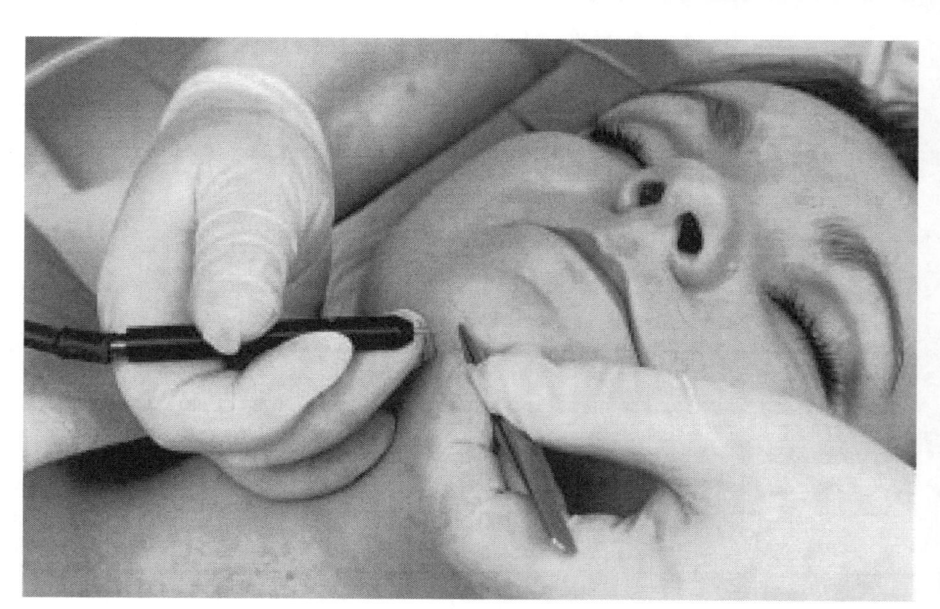

Figure 13.16 Chin, two-handed method, right-handed operator

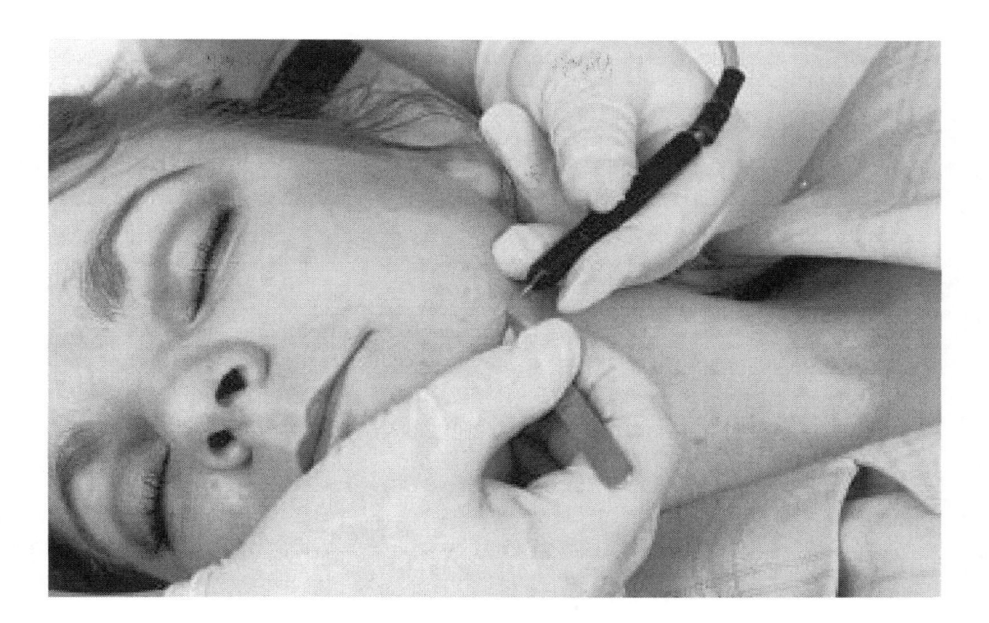

Figure 13.17 Chin, two-handed method, left-handed operator

Review questions

1 What is meant by 'blend'?
2 How does blend epilation work?
3 Why is blend more versatile than either galvanic electrolysis or short-wave diathermy used independently?
4 Describe the action of high-frequency current during blend application.
5 Describe the action of galvanic current during blend application.
6 Why is the correct balance of the two currents important during blend application?
7 Explain the term 'lye'.
8 What is meant by the term 'units of lye'?
9 What determines the settings of the two currents during blend application?
10 Describe the two methods most widely used for establishing the current setting for blend.
11 Name the five variations of blend application.
12 Describe the treatment procedure for blend.
13 How does the body technique work?
14 What is the purpose of the body technique?
15 How does the body technique vary from the general technique?
16 Why is it not advisable to use body technique on the face?

14 Non-permanent methods of hair removal

In the early twenty-first century, hair removal is a multi-million pound business, with both sexes looking towards a long-term or preferably permanent solution to the matter of unwanted hair. Technology moved on at a rapid speed during the 1990s. New methods of hair removal arrived on the scene, many of them claiming permanence, some methods emphasized the fact that the systems did not use needles to achieve results. The media, these days does far more research into products and treatments before going to press. Prospective clients are far better informed with regard to what is available, asking many more questions at the consultation than ever before. It is for this reason that the electrolysist needs to be familiar with treatments available; how they work; and the results likely to be achieved. In others words, the electrolysist must be one step ahead of the client.

Temporary methods

Wax

Depilatory wax is a temporary method of hair removal that lasts 4–6 weeks. Hair is allowed to grow until it is approx. 3–5 mm in length. Warm wax is applied to the area. A paper or fabric strip is pressed firmly onto the wax, and then ripped off very quickly against the hair growth. The hair is torn out of the follicle. This is acceptable for the legs and areas such as the bikini line but not for the face. It must be remembered that facial hair is under hormonal control. When hair is torn out of the follicle the follicle may become distorted, so encouraging ingrown hairs. Blood goes to injured or damaged skin tissue, and, bearing in mind that blood carries oxygen, nutrients and circulating hormones, it is inevitable that the follicle will regenerate to produce a stronger and healthier hair. Fine, downy hairs are also removed and if these follicles are sensitive to circulating hormones they may be stimulated into producing terminal hairs.

Sugaring

Sugaring dates back many centuries – to the days of Anthony and Cleopatra (69–30 BC). It is a mixture of sugar, lemon juice and water, which forms a sticky paste. This is warmed to a temperature of 55–60°C for strip sugar and 50–55°C for hand sugar and then applied to the skin firmly with the fingers, and worked onto the hair (so gripping it), which, in turn, is ripped out of the follicle with a lifting motion. The effects are similar to those of depilatory wax.

Shaving

Shaving removes the hair level with the skin's surface. The end will be blunt, therefore regrowth will feel bristly and rough to the touch. Regrowth occurs within two to three days (the majority of men need to shave on a daily basis). Shaving can cause skin irritation.

Plucking/tweezing

This has exactly the same effect as waxing on the hair follicle. When clients have been plucking hairs on the chin area for some time they are often left with a dark shadow and/or bruising in the area. The skin also becomes desensitized, rough to the touch, with a coarse appearance.

Depilatory creams

These creams are effective for four to five days. The cream is applied in a thick layer where the active ingredients (which are alkaline) dissolve the hair, level with the follicle opening. Due to the alkaline nature of the creams, there is a risk of an allergic reaction occurring in the skin. When this happens the skin becomes inflamed, red and itchy and it may take several days to return to normal. It is essential that a patch test is carried out on the area prior to applying depilatory creams.

Threading

The skilful application of cotton sewing thread held between the thumb and forefinger of each hand achieves hair removal. The thread is then moved backwards and forwards, in a rhythmic motion, with the cotton closing over the hair, which is then pulled out of the follicle. This method is effective when treating eyebrows, but, as with waxing and tweezing, is not recommended for the rest of the face. Threading is practised extensively among the Asian community.

Mechanical epilators

These electric, hand-held devices work by grasping the hair and tearing it out of the follicle by means of a metal coil. The side effects associated with this method of hair removal are ingrown hairs, inflammation and folliculitis. As with waxing, tweezing and threading, mechanical epilators can be painful to use.

Electrical epilators

There are also number of electrical depilation units that are offered as an alternative to conventional electro-epilation. All promote the fact that needles are not used throughout the treatment. Results are achieved by applying a galvanic current, high-frequency current or sound waves through electrodes. With some equipment, the hair is held by tweezers, which are used to pass the current down the hair into the follicle. With other equipment a gel is applied to the skin and the electrodes moved over the area.

The number of machines aimed at permanent hair removal without needles grows annually. It is important that electrolysists keep up to date with these trends so clients' questions can be answered fully.

At present, electrolysis is the *only recognized and proven method of permanent hair removal.*

Long-term hair removal methods

Lasers and intense pulsed light

Lasers and intense pulsed light (IPL) hair removal systems are a recent innovation in the hair removal business. Lasers for hair removal first caught the imagination of the general public when the ruby laser was featured on the TV programme *Tomorrow's World*. The impression gained by many people was that permanent hair removal could be achieved in a single treatment. In reality this does not happen. However, technology is expanding and advancing rapidly owing to the demands of the general public, who require a fast method that is as effective as the long-established and proven method of electrolysis.

There are number of definitions associated with lasers and their application, which the operator needs to be familiar with to achieve the best results.

- **Coherence** – defined as light waves that are spatially and temporally in phase with one another, i.e. light travels in parallel waves with each peak being exactly the same distance apart.
- **Monochromatic** – refers to the ability of the laser energy to emit light at a single colour expressed as wavelength, i.e. blue, orange, yellow, green, red.
- **Collimated** – waves of light are emitted in parallel to one another, without divergence as the beam travels through space, e.g. rows of soldiers moving in an organized manner, parallel to each other. The properties of collimation and coherence permit the laser energy to be accurately focused to a very small beam of light.
- **Wavelength** – different wavelengths have different effects on tissue. The longer the wavelength, the deeper the penetration, therefore it is important to choose the correct wavelength for the target chromophore.
- **Chromophore** – is the term given to the absorbing material, i.e. light-absorbing target. Melanin contained within the hair is the chromophore for intense pulsed light and laser when used for hair removal. Melanin absorbs red light, whereas oxyhaemoglobin in the blood absorbs yellow light.
- **Pulse duration** – refers to the length of time the intense pulsed light or laser is on the tissue. Short pulses of between 10 and 50 nanoseconds induce mechanical damage, whereas longer pulses induce thermal damage. The optimum pulse width for photo-thermal hair removal is between 1 and 50 milliseconds.
- **Fluence** is the amount of energy delivered to a unit area in a single pulse. Continuous lasers are measured in watts, whereas pulsed lasers are measured in joules per cm^2.

What is a laser?

The word 'laser' is an acronym for *light amplification by stimulated emission of radiation.*

A laser is a piece of equipment that produces a monochromatic (single wavelength), collimated (parallel), coherent (the waves are in phase) beam of light. Laser light may be continuous or pulsed, and can be focused on a tiny spot/area, concentrating its energy so it can be used for cutting, welding, in medical surgery and in hair removal.

There are a number of lasers now in use for hair removal. Some are more effective than others. Lasers, when used for hair removal, work at a fixed wavelength of between 694 nm and 1064 nm. The most frequently used lasers for hair removal are:

1. Ruby laser.
2. Alexandrite laser.
3. Long pulsed Nd:YAG.
4. Q switched Nd:YAG.
5. Diode.

The advantages of lasers are that many hairs can be treated in a single session; treatment is rapid, long-term in some instances and cost effective.

Lasers are suitable for treating dark hair on fair skin where there is a definite contrast in melanin content.

Lasers are classed from 1–4. Classes 1, 2 and 3a do not require registration with the Local Health Authority; however, some class 3b and all class 4 lasers require registration. This may well change in the future.

The disadvantages of lasers are that they can be extremely painful; there may be adverse skin reaction; and they are more or less limited in their application to treating dark hairs on light skin. Lasers in general do not have the versatility in application of intense pulsed light. Side effects can be hyper- and hypopigmentation of the skin, and blistering can also occur.

What is intense pulsed light?

Intense pulsed light (IPL) systems use a flash lamp to emit white light, of variable wavelengths, which contains the visible rays of the spectrum. The wavelengths can be adjusted to suit the hair and skin type. The light is non-ionizing and non-invasive. IPL works by selective photothermolysis.

Selective thermolysis can be defined as:

- *Selective* – targets specific sites, e.g. melanin in hair and follicle.
- *Photo* – light.
- *Thermolysis* – heat.

Selective thermolysis destroys the target by means of heat produced by the application of light.

The aim of IPL is to heat the lower third of the follicle to a temperature high enough to destroy the germinative cells without causing damage to the surrounding tissue. IPL can be varied to treat different skin types and hair colour. However, to achieve long-term hair removal, the colour of the hair must be darker than that of the skin.

As with electrolysis, the treatment is only as good as the operator. Side effects of intense pulsed light are similar to those of laser. It is essential that treatment with IPL is not applied to skin that has recently been exposed to sunlight or to a client who is taking photosensitive medication.

At the beginning of the twenty-first century, intense pulsed light and laser technologies are in their infancy and as yet it is too early to claim permanent hair removal as opposed to long-term hair removal. However, technology is advancing rapidly and it will be interesting to watch developments. Certainly the electrolysist needs to keep abreast of developments to be able to answer clients' questions in an informed manner.

Comparison of lasers/IPL and electrolysis

IPL/laser

- Fast, effective.
- Large areas – backs legs, arms, chest.
- Frequency: 4–12 weeks.
- Up to 85% clearance over 12 months.
- 3–10 treatments.
- Cost effective.
- Improved quality of life.
- Long-term hair removal.

Electrolysis

- Effective, proven, 1875
- Slow but sure
- Ideal small areas
- Frequency: from weekly
- Impossible to state number of treatments
- Results dependent on skill of electrolysist
- Recognized as a *permanent method* of hair removal by the British Medical Association.

Conclusion

There are a number of temporary methods of hair removal, all of which give only a short-term solution to the problem of unwanted hair. Regrowth from shaving is blunt and uncomfortable, with hairs growing in several different directions. Sugaring, waxing and plucking all encourage stronger regrowth on the face and a tendency to develop ingrown hairs. Depilatory creams and preparations are messy to apply and can result in contact dermatitis.

At the beginning of the twenty-first century, lasers and intense pulsed light systems are at the early stages of development. Whilst showing positive results for long-term hair removal, it is far too early to determine whether or not these results are going to be permanent.

On balance, although electrical epilation by galvanic, short-wave diathermy or blend is slow, in the right hands it is both safe and effective. To date, it is the only method of hair removal that has been recognized as permanent by the British Medical Association.

Review questions

1. Why is it not advisable to wax facial hair?
2. Name two disadvantages of depilatory creams.
3. State the disadvantages of plucking or tweezing facial hair.
4. List the temporary methods of hair removal.
5. Why are laser and intense pulsed light treatments referred to as 'long-term hair removal' rather than 'permanent hair removal'?
6. What is a laser?
7. Define 'selective photo-thermolysis'.
8. State the differences between a laser and intense pulsed light.
9. Compare lasers and intense pulsed light with electrolysis.

15 Consultation

The consultation provides a golden opportunity to form a link between the client and the electrolysist. In many instances a great deal of courage is required for the client to make the initial appointment, and further courage is needed for him/her to walk through the clinic door. The electrolysist's approach at this stage will either encourage or discourage the potential client.

What are clients looking for during the initial contact with the electrolysist? Usually they are seeking a professional approach which is warm and welcoming – one that is not patronizing or flippant, or which makes them feel abnormal or inferior. Questions may arise relating to the professional qualifications of the electrolysist; knowledge of the subject; frequency and length of treatments; time taken to clear the problem permanently; methods of dealing with the hair growth between treatments, cost involved; approach to hygiene; and the use of sterile disposable needles.

Having considered the advantages to the client, some thought should be given to the benefits of the consultation to the electrolysist. Valuable information can be obtained at this stage that will help the electrolysist plan an effective course of treatment. Several factors need to be considered before a decision on treatment is reached.

Hair

- Type of hair growth: fine, vellus, terminal, curly, straight.
- Density and site of hair growth: chin, upper lip, body, bikini area.
- Previous methods of hair removal: waxing, plucking, shaving, depilatory creams, cutting.

Previous hair growth control and the frequency with which a particular method has been used must be discussed to establish whether follicles may have become distorted as a result of waxing, sugaring or plucking. If this is the case, then treatment with the blend technique may be more effective than short-wave diathermy, due to the inability to place the needle accurately at the base of the follicle. When the client has been plucking regularly she often does not realize the full extent of the problem, which may come as a shock once the tweezers have been banished!

Scarring may be present from electrical epilation received elsewhere. If so, type and extent should be noted, as should the length of time during which scarring has been evident.

Skin

Several points need to be looked at here: whether the skin is fine in texture or coarse; dry; oily or sensitive; the healing rate; and the presence of acne, infection, eczema, psoriasis or other skin conditions in the area to be treated. Does the client have any allergies to certain preparations or metals? The skin can also be an indicator of underlying medical conditions (see below).

Medical history

At this stage a detailed medical history should be taken, which must include the following:

- Possible contraindications such as asthma and emphysema.
- Prescribed medications: steroids, anti-depressants, hormones, hormone replacement therapy; contraceptive pill; drugs for the control of epilepsy.
- Pregnancy: stage and any complications that may have arisen during the pregnancy.
- Hepatitis B: date of illness, prescribed drugs (notification of the proposed treatment to the general practitioner).
- Hepatitis C.
- Diabetes: whether it is controlled by diet, tablets or insulin injections. Healing rate, tendency to bruise and sensitivity of skin should be noted.
- Epilepsy: severity and frequency of fits. Medication used. Ascertain whether or not the condition is controlled.
- Endocrine and gynaecological problems: polycystic ovary syndrome, menstrual irregularities, menopause, endometriosis.
- Organ transplant, i.e. lungs, kidney; nature; date; and medication.
- Surgery: date and nature.

There are many reasons for obtaining a medical history. Most importantly, the information enables the electrolysist to assess whether or not the treatment is right for the client. Conditions such as diabetes require liaison with the client's GP prior to commencing treatment. Medical treatment is required for conditions such as polycystic ovary syndrome, since without this, complete elimination of the hair growth is not possible.

During the menopause it is possible that hair growth will increase due to the change in the hormone balance. Some follicles become sensitive to these circulating hormones in the blood, and hair growth is stimulated. This can and does cause embarrassment.

A woman can become more sensitive with a lowered pain threshold immediately before and during menstruation, therefore the treatment will often be painful at this time. The healing rate of the skin may also be slower.

During pregnancy there is an alteration in hormone levels and in a number of instances hair growth increases. This growth is often temporary and disappears after the birth of the baby without any treatment at all. There is no known reason why treatment of an existing problem should stop during the pregnancy but it is advisable to notify the client's GP of the treatment details. Blend treatment should not be given during pregnancy.

Certain medications stimulate hair growth, e.g. steroids and some forms of hormone replacement, whereas others increase the risk of pigmentation, e.g. the contraceptive pill and those hormones used in large doses for transsexual clients.

Diabetic conditions require liaison with the GP and consideration should be given to the fact that skin is slower to heal, therefore treatment sessions should be spaced further apart. The pain threshold is also lower, particularly before a meal when the blood sugar levels will be low, therefore the timing of appointments needs to be carefully planned.

Also essential to the consultation is an examination of the skin, preferably with the aid of a magnifying lamp. The condition and colour of the skin often gives an indication as to the health of the client, for example, the colour may be high; there may be dilated capillaries, veins may be red or blue. If redness is present in a butterfly appearance across nose and cheeks, rosacea may be indicated. Should the colour be blue, there could be a respiratory or heart problem present.

A dull grey appearance could indicate a smoker – oxygen supply to the skin will be affected due to the constricting effect of nicotine on the small blood vessels, this in turn, will affect the healing rate of the skin. A yellow tinge to the skin could indicate a problem either with the gall bladder or the liver – with liver diseases there is often a tendency to develop spider naevi. A thin skin with a tendency to redness and sensitivity could be the result of steroid application.

Next, look at the condition and type of the skin, e.g. is the skin oily with comedones; is it dry through lack of sebum; or dehydrated through lack of moisture? Does the skin bruise easily? Are there any pigmentation marks? If so, query the possible causes, e.g. medication, contraceptive pill. Is the texture coarse, thick, thin or fine? Are there signs of conditions such as eczema, psoriasis, acne?

If scars are present from previous electrical epilation treatment it is essential these be pointed out to the client, as tactfully as possible, during the consultation. This will prevent any confusion at a later date as to the length of time the scars have been evident. In other words, it will protect the electrolysist who gives further treatment.

Where there is any doubt as to the client's suitability for treatment, permission should be obtained to contact her GP. A letter such as that shown in Figure 15.1 should be sent, giving the GP details of the planned treatment.

There are rare occasions when a prospective client will be reluctant to give the electrolysist information relating to medical history and previous treatment during a consultation. This type of client can be difficult to treat. Explain to the client why the information is needed. Quite often, by going through the different questions with an explanation of why they are being asked, the client will respond with a negative or positive answer. For example, 'that doesn't apply to me', or 'I am not taking hormone replacement or medication'. In this way it is sometimes possible to gain the relevant information. When the prospective client refuses to answer questions relating to medical history and will not allow you to contact her GP, then it would not be wise to proceed with the treatment.

Alternatively, continue the consultation without pursuing the matter of the medical history. Give a detailed explanation of electrolysis and how it works and then give the prospective client a leaflet that explains what electrolysis is and how it works (these are available from specialist professional associations). Suggest the prospective client goes home to give the matter further thought before undertaking treatment.

At this stage a short explanation of the hair-growth cycle helps the client to understand what is happening below the skin's surface and the importance of attending for treatment on a regular basis. The removal of a few

Your Ref:

Doctor's name
Doctor's address

Dear Dr ...

Re: Patient's full name
Patient's address

Miss/Mrs ... has been to see me in connection with removal of hair by (*insert method of treatment, e.g. SWD or Blend*) from the (*insert area*).

Would you please confirm that this patient is suitable for treatment? Also could you advise me of any medication she is taking that may have an influence on hair growth or the proposed treatment.

Yours sincerely,

Electrolysist's name
Qualifications

There is no reason why

____

should not receive electro-epilation for hair removal.

Signed______________________

Dated______________________

Figure 15.1 Pro forma letter suitable for sending to the client's GP

hairs will enable the client to see how treatment feels, while giving the electrolysist an opportunity to assess the skin's reaction.

A simple description of how electrical epilation works will give the client an insight as to why it is not possible to guarantee the destruction of any one hair permanently after a single treatment.

The electrolysist should now be able to advise the client on the frequency of appointments, length of each session and cost per visit. It is not usually possible, or advisable, to give a time of completion concerning the permanent elimination of the problem.

Assessing the client

Is the client calm and confident or nervous, agitated, embarrassed, or on the defensive or attack? What is the client's lifestyle? Is she prepared to follow instructions for aftercare and management of hair growth between treatments?

During the consultation the electrolysist should take the opportunity to enquire how the client came to hear about the clinic/electrolysist. Was it by recommendation; through a professional association; advertising; or passing by?

When conducted thoroughly, the consultation gives the electrolysist a comprehensive picture of the new client and her needs. Any contraindications that may be present should be observed at this time, and an explanation given to the client as to why treatment with electrical epilation is not recommended. This client should be placed under no obligation to proceed with treatment should she not wish to do so. When a consultation has been conducted in the right manner with the correct professional approach, there are very few clients who leave the clinic without making a further appointment.

The effects of unwanted hair growth on a client's psychology

The presence of unwanted hair growth usually has a demoralizing effect on a woman, resulting in loss of self-confidence. The individual may show signs of being inhibited, shy, aggressive, argumentative or defensive to mask her feelings of embarrassment. The electrolysist should be able to read the body language displayed by the prospective client to assess the best method of approach. For many individuals a great deal of courage is needed to walk through the clinic door at the scheduled time. The electrolysist's manner and approach at the initial meeting is of vital importance.

The author never fails to be pleasantly surprised by the personality changes that take place in many clients once the course of treatment has been started. Shy, inhibited individuals become more outgoing and interested in life in general. Aggressive, argumentative and defensive traits disappear. The client becomes more confident and relaxed and the hair-growth problem becomes less of a dominant factor in her life. A statement made by many clients is 'electro-epilation has changed my life. I feel like a different person and so much happier'. These changes are often due to the sympathy, understanding and professional manner of the electrolysist, in combination with the successful results of electro-epilation treatment.

Client's expectations of treatment

During the initial consultation it is helpful to find out what the client's expectations of treatment are. In many instances the prospective client is under the impression that hair removal will be permanent after the first electrolysis treatment. Often they have read misleading advertisements or read articles in magazines or the local papers.

It is at this stage the prospective client should be given a clear understanding as to the role of electrolysis and that for the best results a commitment should be made by the client to follow the treatment plan and advice on hair control recommended by the electrolysist.

It is important to ensure the client understands that the hair growth did not occur overnight, therefore it will not disappear overnight. Electrolysis is a progressive treatment that will achieve results in the end. An explanation of the hair growth cycle during the consultation will help the client to have an understanding of why there sometimes appears to be instant regrowth after treatment and why the problem cannot be cleared overnight.

Consultation/ record cards

During the consultation the electrolysist should prepare and complete a record card for the client. This should contain all relevant details relating to each client. Information to be recorded includes:

- Medical history and prescribed medications.
- Menstrual cycle, irregularities and number of pregnancies (which may not be the same as number of children).
- Possible cause of hair growth, e.g. polycystic ovaries or hereditary disposition.
- Previous methods of hair removal and effect on skin and hair growth.
- Length and frequency of previous electro-epilation treatments.
- Presence of pigmentation marks or scars.
- Skin type and healing rate.

Once the client has started a course of treatment the following details should be recorded at the time:

- Date and length of treatment session.
- Method of electro-epilation used, together with intensity of currents.
- Needle size, texture of the hair and area treated.
- Appearance of the skin and any reaction to treatment noted.

Accurately recorded details of treatment enable the electrolysist to monitor the progress of treatment together with alterations in hair growth patterns and distribution. This will enable the electrolysist to assess future treatment requirements according to the client's needs. An example of a record card is shown in Figure 15.2.

Treatment of young girls

Some very young girls are nowadays plagued with the presence of dark hair on the upper lip. This is often seen in Asian and Mediterranean races where the hair colouring is naturally dark and there is a natural predisposition to hypertrichosis. The question is often asked 'At what age can young girls be given treatment?' This is a tricky one. Treatment given to anyone under the age of 18 years, without the consent of a guardian or parent, is technically an assault. That aside, should electrical epilation be given prior to puberty?

With the increasing incidence of 'bullying' at school by other youngsters it is often necessary to weigh up the advantages of giving treatment to a minor rather than waiting until the hormone levels have settled down. For many young girls the psychological effects of dark hair can be detrimental to the future emotional development of the child. Two illustrative case histories are included in Chapter 25.

INSTITUTE OF ELECTROLYSIS CONSULTATION CARD

Date of Consultation: Consulting Therapist:

Client's Name: Male/Female

MEDICAL HISTORY

Have you or are you currently suffering from the following conditions:

Mark every box	YES	NO	DETAILS/DATES/MEDICATION
Heart Disorders/Angina			
Pacemaker fitted			
Haemophilia			
Hepatitis			
HIV+			
High Blood Pressure			
Low Blood Pressure			
Blood Thinning medication/warfarin/asprin			
Anti-inflammatory medication			
Varicose Veins/Thread Veins			
Dilated Capillaries			
Likely to Bruise/or present			
Antidepressant medication			
Steroids-Topical/Oral			
The Pill/HRT			
Diabetes			
Psoriasis			
Eczema			
Herpes Simplex			
Skin conditions i.e.fungal			
Allergic Reactions			
Recent Scar tissue			
Metal Implants/DentalWork			
Epilepsy			
Psychiatric Disorders			
Recent Operations			
Recent Hospitalisation			
Pregnancy			
Thyroid/Ovary disorders			
Infection/disease			
Fever/ Inflammation			
Recent Anaesthetic			
Respiratory Conditions			

SKIN & HAIR DIAGNOSIS

Areas to be treated:

Mark each box	**YES**	**NO**	**COMMENTS**
SKIN TYPE IN AREA TREATED			
Greasy			
Dry			
Sensitive			
Young			
Mature			
Sun Exposure			
Pigmentation			
Moist/Perspires			
Loose Skin			
Firm			
Pustules			
Inflammation			
Moles/Papillomas etc.			
Laser treated areas			
HAIR IN AREA TREATED			
Terminal/Vellus/mixed			
Sparse/Dense			
Pigmented / non-pigmented			
In Grown Hairs			
How long has the hair been present?			
Suspected hair cause			
How have you treated the hair until now?			
How often do you treat it?			
How quickly does the hair grow?			
Have you experienced Epilation before?			
How does your skin react to touch/ treatments/products?			
How well does your skin heal?			
Have you experienced any abrasive/peel type treatments recently?			

THE INFORMATION I HAVE GIVEN ABOVE IS TRUE TO THE BEST OF MY KNOWLEDGE, AND I AM NOT KNOWINGLY WITHHOLDING INFORMATION REQUESTED OF ME. ( )
I HAVE RECEIVED A CONSULTATION RELEVANT TO THE ABOVE STATED TREATMENT. ( )
I WILL FOLLOW THE VERBAL AND WRITTEN AFTERCARE INSTRUCTIONS AND ADVICE WHICH HAS BEEN GIVEN TO ME.( ) Signature:

Figure 15.2 Example of a consultation card

Review questions

1. List the advantages of the initial consultation to the client.
2. List the advantages of the initial consultation to the electrolysist.
3. Why is it necessary to discuss previous hair growth control?
4. What details should be taken relating to medical history?
5. State the advantage of explaining the hair growth cycle to the client.
6. What should the electrolysist look for when examining the client's skin?
7. State the information that can be obtained from assessing the client during a consultation.
8. Explain the psychological effect on a client of unwanted hair growth.
9. State the information that should be entered onto the client's record card.
10. Why is it necessary to keep accurate records of treatment details?

16 Contraindications

There are a number of conditions that are either contraindicative – i.e. show that treatment is inadvisable or dangerous – or which require advice from the client's doctor before beginning electro-epilation.

- Asthma: defined as 'a condition characterized by transient narrowing of the smaller airways'. During an asthmatic attack the patient experiences great difficulty in breathing. An attack may be triggered by anxiety or stress. Should the client's GP agree to electro-epilation treatment, particular attention should be given to positioning during the session.
- Dermagraphic skin condition: congenital sensitivity to any form of friction on the skin. Swelling appears shortly after treatment and may last up to 24 hours. No long-term adverse effects, but the decision concerning continuation of treatment should be the client's.
- Dermatitis/eczema in the treatment area: increased sensitivity, skin irritated, and often a build-up of dry skin blocks opening to follicle thereby hindering insertion.
- Fungal infections, e.g. tinea: risk of transmitting infection.
- Bacterial infections, e.g. impetigo: risk of transmitting infection.
- Viral infections, e.g. herpes simplex, herpes zoster: risk of cross-infection.
- Heart problems/circulatory disorders requiring medical treatment: advisable to consult with GP.
- Haemorrhage/bruising: disturbed blood supply interferes with healing process. Heating effect of short-wave diathermy causes blood vessels to dilate.
- Hypertension/anxiety, stress: nerve endings highly sensitive, client unable to relax therefore treatment more painful. Insertion to follicle hindered. Risk of scarring due to client pulling away during current application.
- Loss of skin sensation: inability to sense when current is too high, which could result in over-treatment of area.
- Metal plate: concentration of high-frequency field causes overheating in tissues.
- Pre-malignant/malignant lesions: possible stimulation of metabolism due to increase in temperature could accelerate growth.
- Hepatitis C.

An additional consideration when using galvanic or blend treatment is the presence of excessive dental fillings which often give rise to a metallic taste in the mouth and which may be unacceptable to the client.

Conditions that require liaison with the GP prior to treatment

- Epilepsy: electrical impulses to the brain may be disturbed, which could result in a fit. Short-wave diathermy only should be given, particularly if anxiety is present.
- Vascular disorders requiring anticoagulant drugs: will affect the coagulation of the blood supply at the base of the follicle.

- Hepatitis B: strict attention to hygiene and the use of disposable sterile needles is essential. The clotting mechanism is often affected, and the healing rate of the skin is inhibited. The client bruises easily.
- Naevi/moles: hairs growing out of moles must be referred to a general practitioner for checks on possible malignancy.
- Diabetes: slow to heal; low pain threshold; shorter treatments; longer healing gaps; increased length of time between treatment sessions.
- Endocrine disorders: several endocrine disorders such as polycystic ovary syndrome result in increased hair growth in a male pattern. It is essential that correct medical treatment is carried out, which may help in the reduction of unwanted hair. However, it is worth remembering that electrical epilation in conjunction with medical treatment will speed up the final result.
- Emphysema: some drugs used in the control of emphysema can lead to increased hair growth. In extreme cases it may only be possible to keep the hair growth under control rather than eliminate it. However, it must be remembered that psychologically electrical epilation will be of great value to the client. Great care and a detailed explanation of the cause of hair growth in this situation must be given at the time of the consultation. The client must be fully aware of the process involved to avoid unrealistic results with regard to length of time and degree of hair removal. Consideration must be given to the positioning of the client during treatment due to breathing difficulties.
- Steroids can often be instrumental in encouraging hair growth.
- Hiatus hernia: positioning of the client during treatment is of the utmost importance. The client should be raised into a sitting or semi-sitting position to avoid discomfort.

Review questions

1. Define the term contraindication.
2. Why are the following termed contraindications to electrical epilation?
   (a) asthma
   (b) herpes simplex
   (c) loss of skin sensation.
3. Explain why it is necessary to liaise with the client's doctor when the following conditions exist:
   (a) epilepsy
   (b) hairy moles
   (c) endocrine disorders.
4. What complications arise as a result of emphysema?

17 Practical application of electrical epilation

The question is often asked 'What is electrical epilation and why is it necessary?' Quite simply, electrical epilation is the permanent removal of unwanted hair. This is brought about by the destruction of the lower third of the follicle, through the application of an electrical current. The current used may be radio or high frequency, direct or galvanic, or a combination of the two currents, referred to as *blend*.

Any person who has a problem with, or is embarrassed by, hair growth could well benefit from electro-epilation. It must be remembered that some problems occur as a result of an endocrine disorder, which could require medical treatment prior to electro-epilation. Hair growth is a problem which affects any age range or social background.

Key points in treatment

The points to be considered when giving a treatment are:

1. Operator's posture and position during treatment.
2. Positioning of the client.
3. Positioning of the equipment.
4. Selection of the needle size.
5. Preparation of the area to be treated.
6. Technique.
7. Rhythm and continuity.
8. Accuracy of probing.
9. Current adjustment.
10. Aftercare and home care advice.

Should any one of the above not be carried out correctly there will be an adverse effect on other areas, which in turn will have a detrimental effect on treatment.

Operator's posture

The correct posture is one that can be maintained with ease throughout the working day. There should be no undue fatigue or strain. The position of the back, shoulder, arm, wrist and hand should be considered in relation to the area being treated.

The back should be straight, with the body tilted slightly forward from the hips when necessary. The shoulders should be relaxed; muscles should not be tensed or raised towards the ears. When lifting or raising the arm, the shoulder should remain relaxed.

Arm, wrist and hand are kept in alignment with the follicle and direction of the hair growth. This will help in achieving accurate insertions.

Incorrect posture will result in fatigue, backache and tightening of the shoulder muscles. This in turn could lead to a number of problems, including, in the long term, a frozen shoulder and/or consistent lower back problems, and in the short term severe headaches and/or painful neck and shoulders.

Standing for any length of time is not to be recommended. This position is tiring, prevents use of the footswitch and affects balance, which in turn will affect probing technique. It must be remembered that at times it will be necessary to stand for clients who need to be kept in an upright position, for example, those with respiratory problems such as emphysema, or conditions such as hernia or vertigo.

Leaning on the client during treatment can be embarrassing, uncomfortable and distasteful to the client.

Operator's position

The correct working position during treatment should enable the operator to:

1 Maintain correct posture.
2 Align the probe to the direction of the hair growth, so aiding entry of the needle into the follicle.
3 Reach the area requiring treatment easily.

To achieve this the operator must work on the *correct* side of the treatment couch. Right-handed operators work on the left side of the couch and left-handed operators work on the right side of the couch.

The operator's working position can be enhanced by the use of a stool or chair that is adjustable in height.

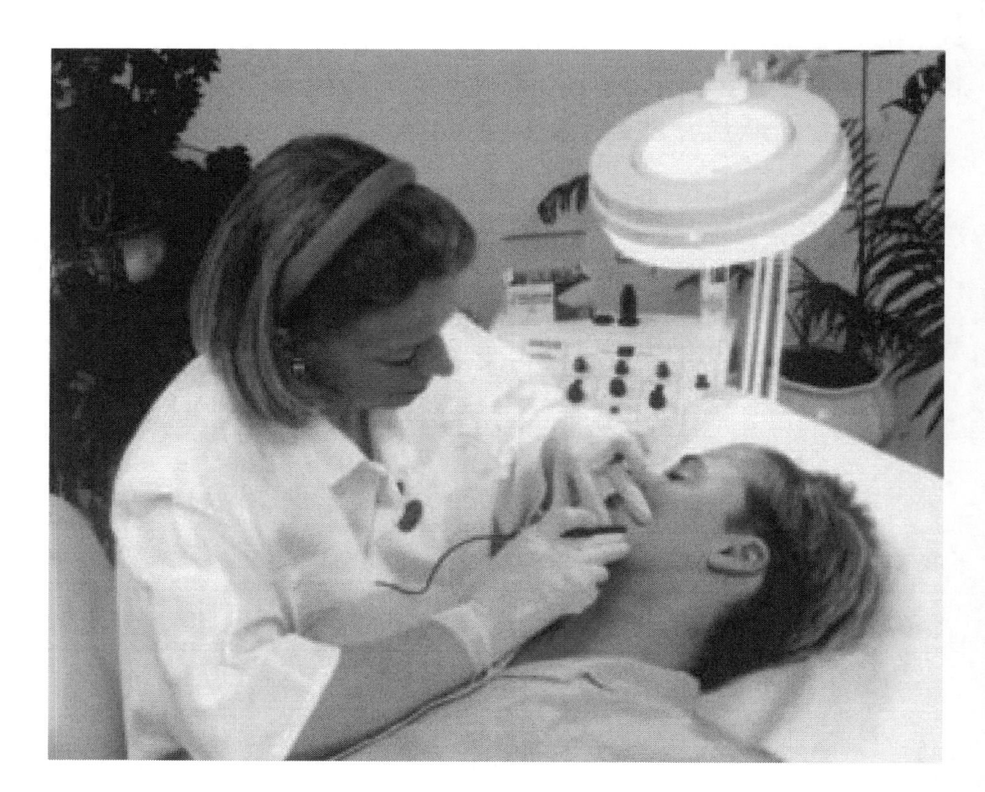

Figure 17.1 Correct operator's position (right-handed)

Assessment of the client prior to and during the treatment

It is essential the operator observes the client carefully both before and during the treatment. The client's mood or sensitivity to treatment varies from one visit to another. Observe whether clients are relaxed or tense when they enter the treatment room, or whether they are tired or make reference to headaches or health problems etc.

During the treatment observe skin reaction, client's response and pain tolerance. Should the client tense during treatment, the current intensity should be reduced. When the client finds the treatment uncomfortable it is better to reduce the intensity and apply the current for a little longer.

Positioning of the client

Clients should be positioned so they are comfortable and relaxed, and that undue embarrassment is avoided. At the same time the area requiring treatment should be easily accessible to the operator. Where there is a problem such as hiatus hernia, or where respiratory difficulties are present, it may be necessary to raise the client into a sitting position. When working on the upper lip it is important the client's breathing is not interfered with in any way.

Elderly clients and those with walking difficulties due to conditions such as arthritis and hip replacements may not be able to get onto the treatment couch easily. This should be taken into consideration and treatment procedure adjusted accordingly.

Clients in wheelchairs may need to receive treatment in their chair. In this instance it will be necessary to reposition the equipment and operator's stool. The operator should ensure it is still possible to insert the needle easily and without causing skin damage.

Preparation of electrolysist's hands prior to treatment

See Chapter 19.

Positioning of equipment

The lamp should be placed so that the area is clearly lit, without causing shadows. When magnification is required, the lens should be parallel to the area to avoid distortion of hairs. When positioned correctly, the operator should be able to move hands freely without knocking the lens. With fine vellus hairs it is often necessary to angle the light to the side, which will make the hairs easier to see.

The machine and trolley should be placed so both are easily accessible during treatment. This will enable the operator to:

1 Adjust the current when necessary without losing continuity.
2 See the machine easily.
3 Reach accessories such as forceps, cotton wool and antiseptic easily.

Leads and cables from the equipment should be placed so the risk of either client or operator tripping over them is minimized.

Needle selection

The type and texture of the hair to be treated should be examined closely. The diameter of the needle chosen should match the diameter of the hair. This ensures the needle tip is able to distribute sufficient current to the base of the follicle. A needle that is too large in diameter will stretch the follicle

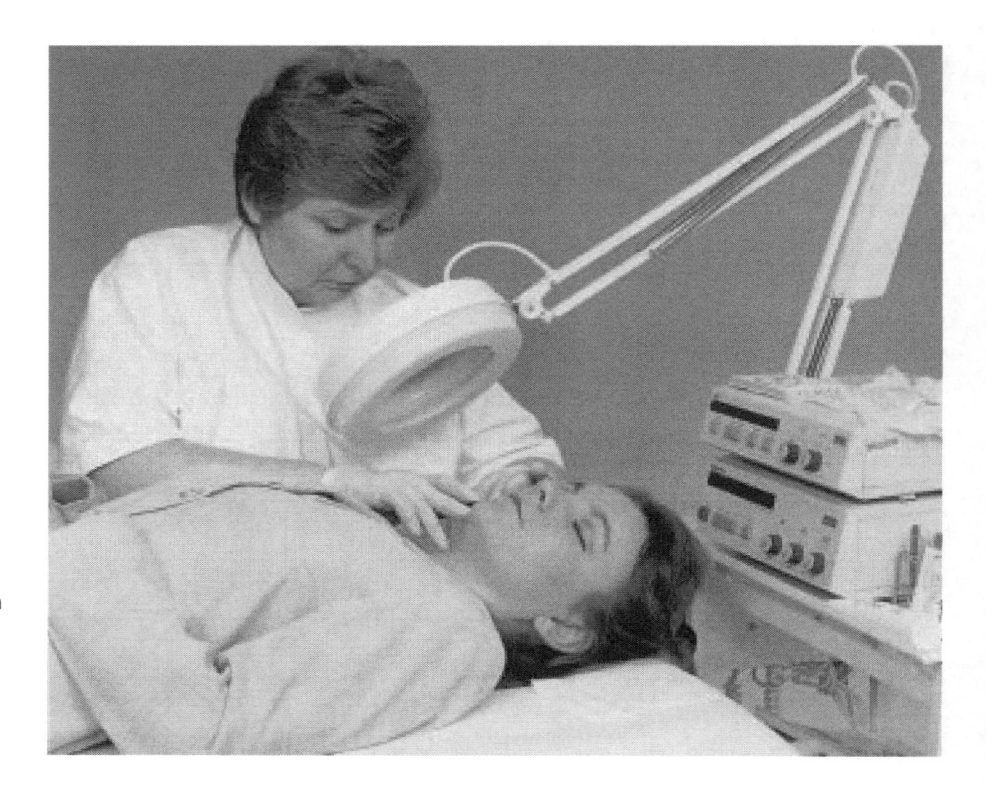

Figure 17.2a Position of right-handed operator in relation to couch, client and equipment when treating the face

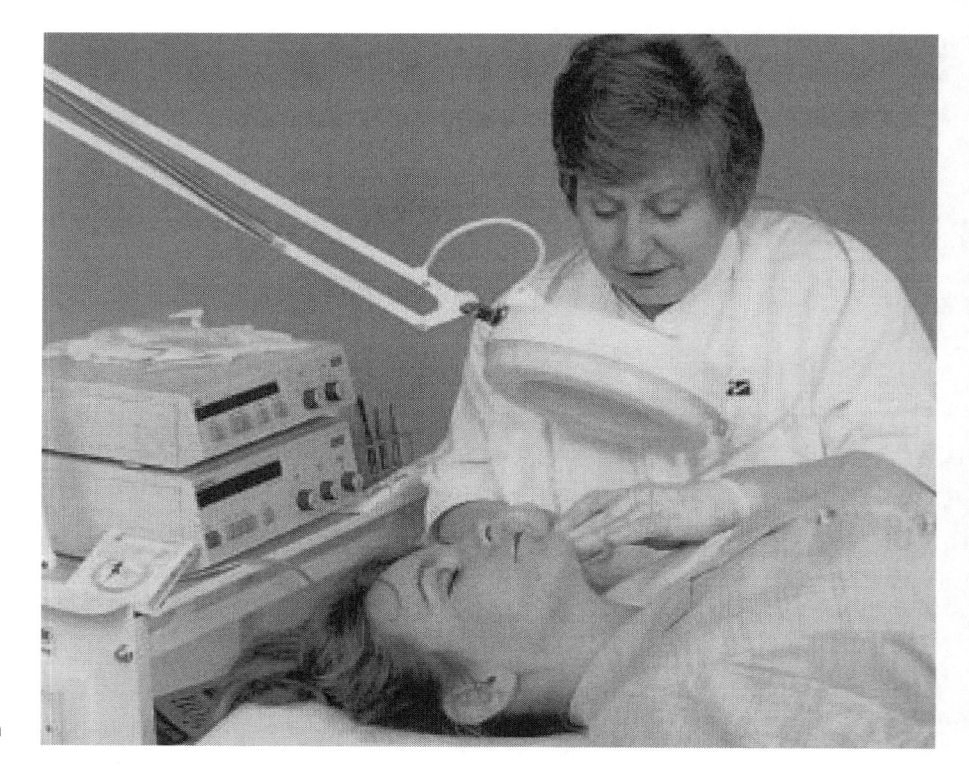

Figure 17.2b Position of left-handed operator in relation to couch, client and equipment when treating the face. Right-handed operators should be seated with the right side of their body next to the couch

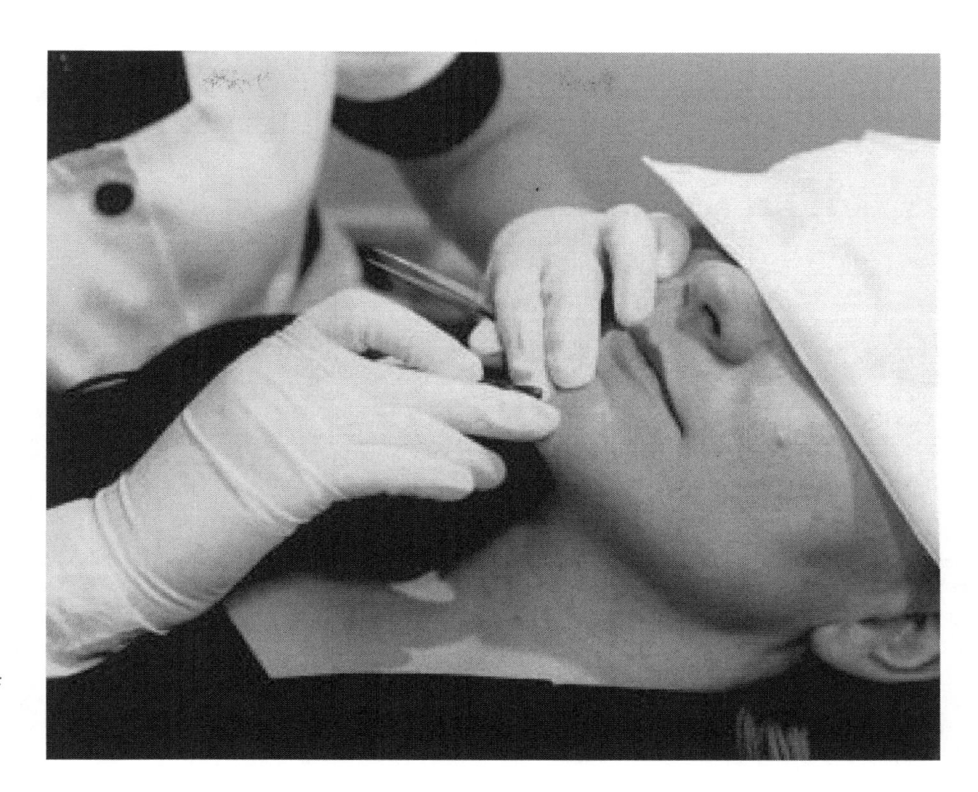

Figure 17.2c Point of chin. Right-handed operator

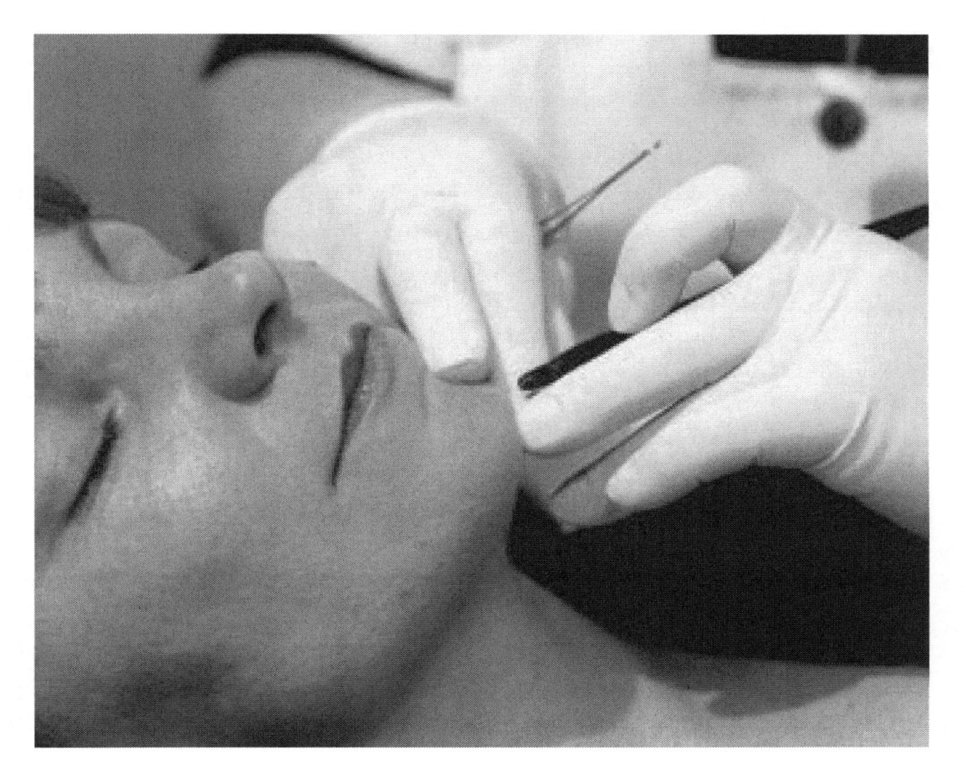

Figure 17.2d Point of chin. Left-handed operator

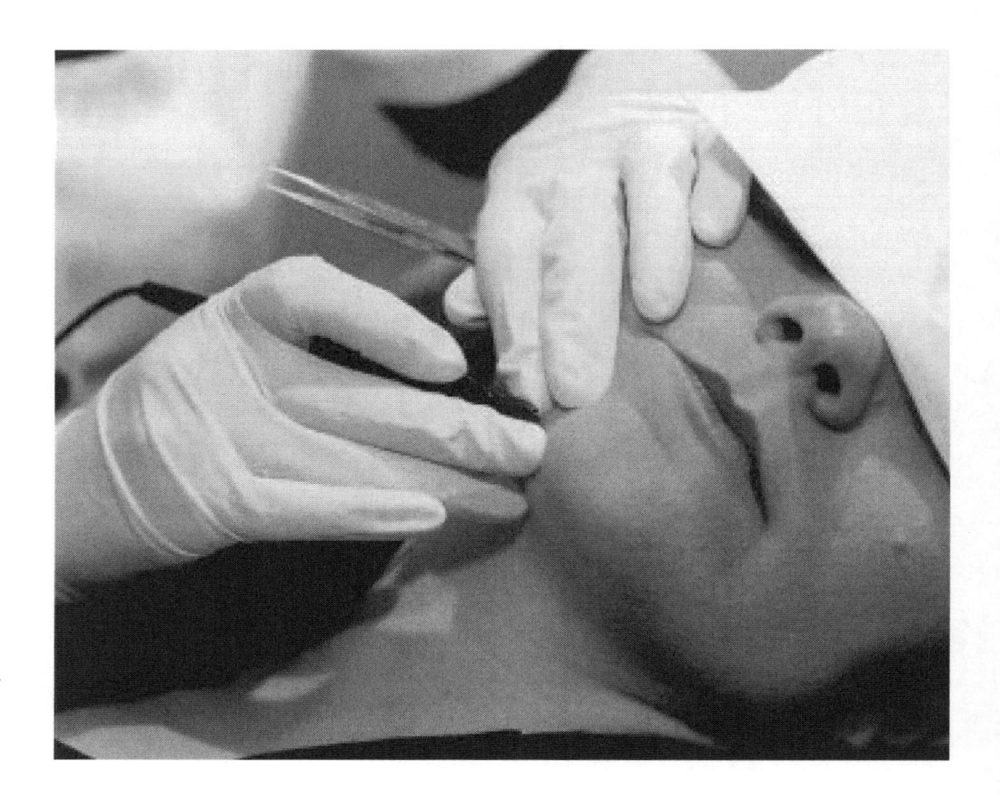

Figure 17.2e Side of chin. Right-handed operator

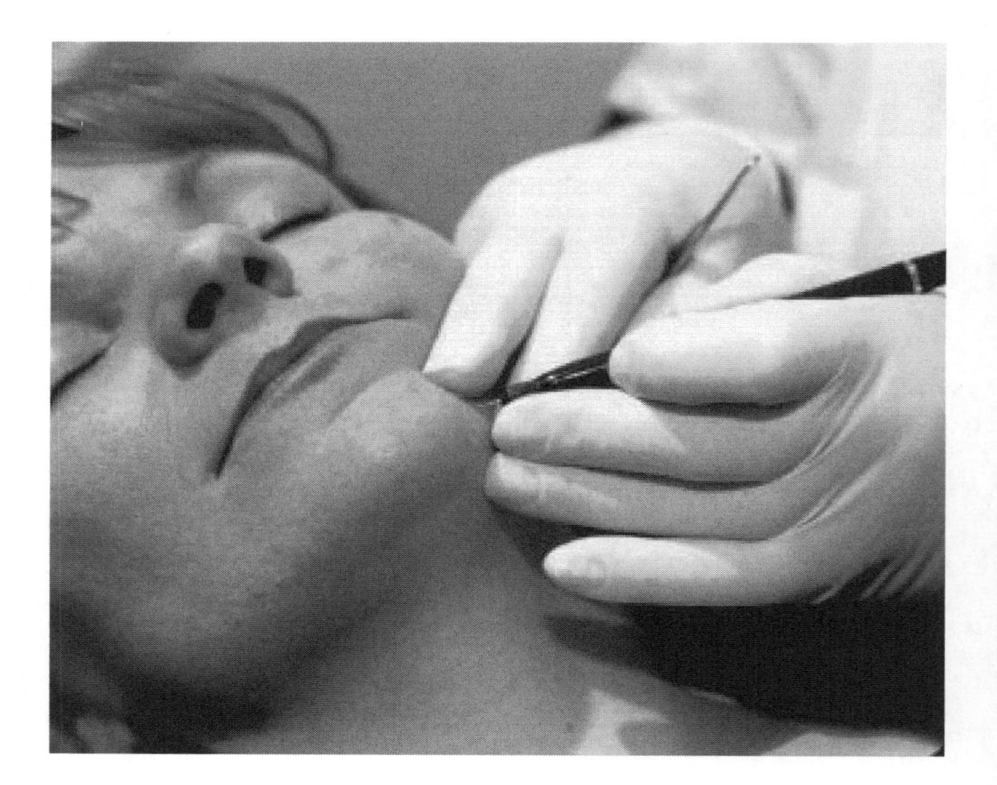

Figure 17.2f Side of chin. Left-handed operator

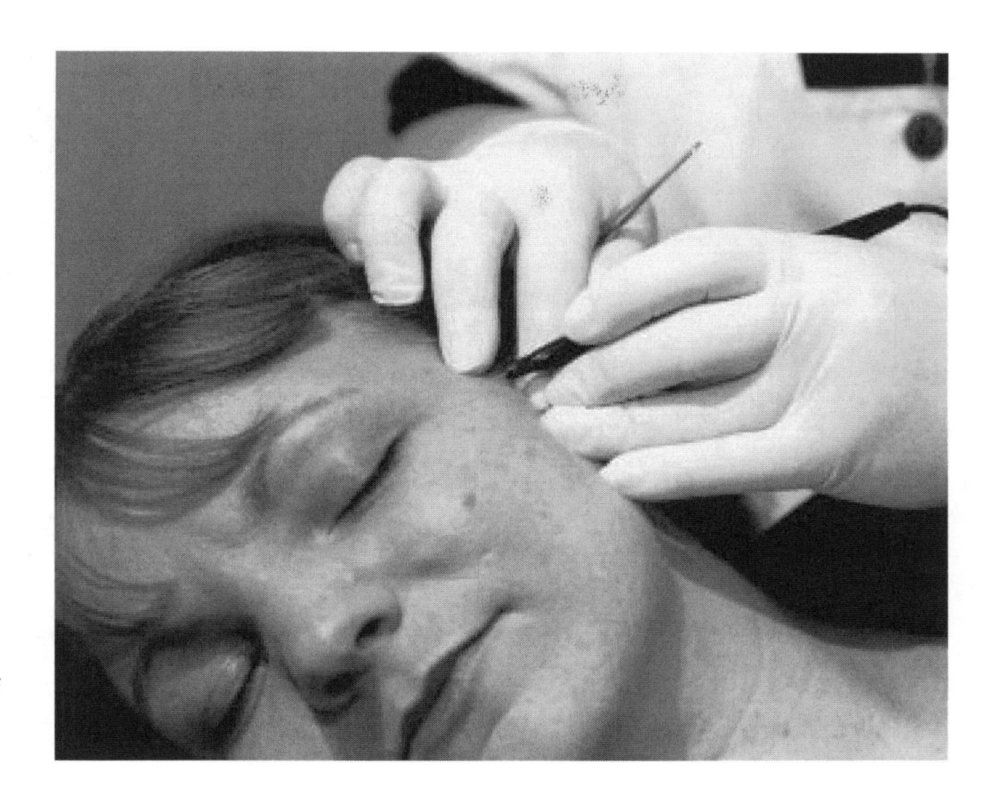

Figure 17.2g Side of face. Left-handed operator

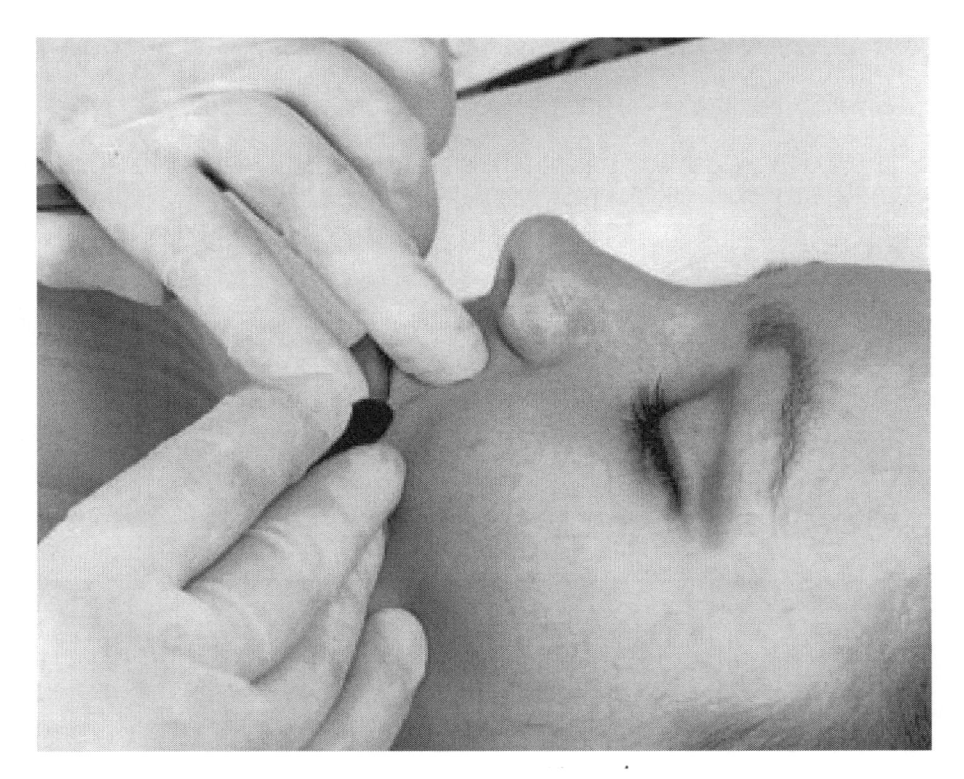

Figure 17.2h Upper lip. Right-handed operator

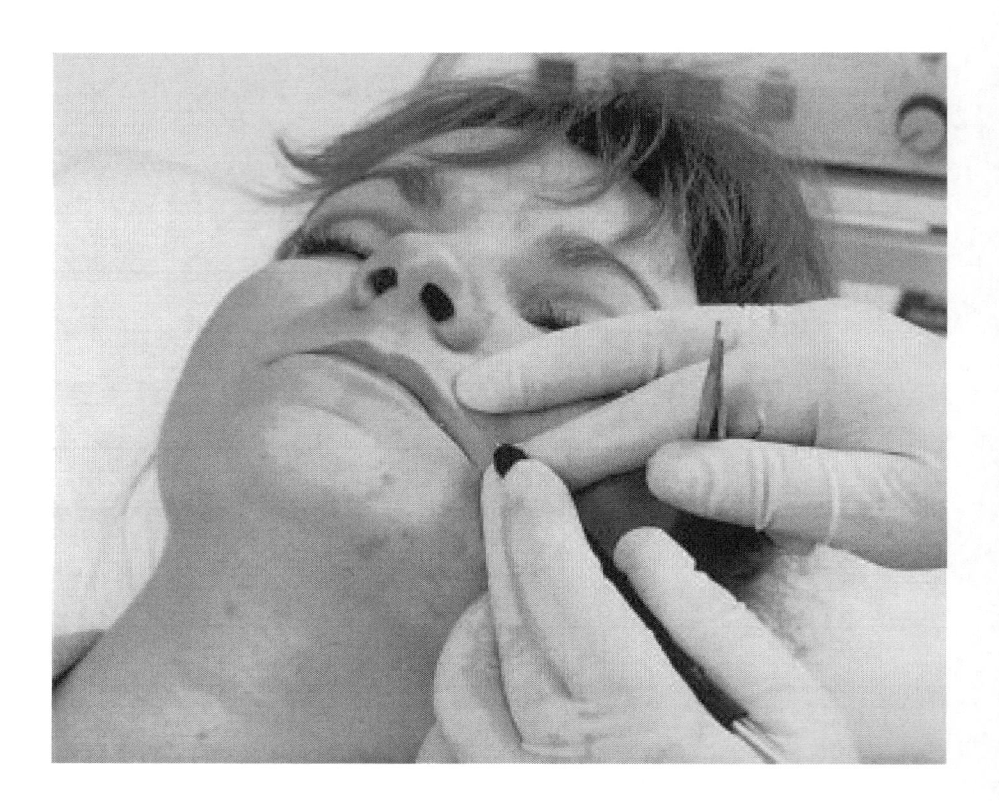

Figure 17.2i Upper lip. Left-handed operator

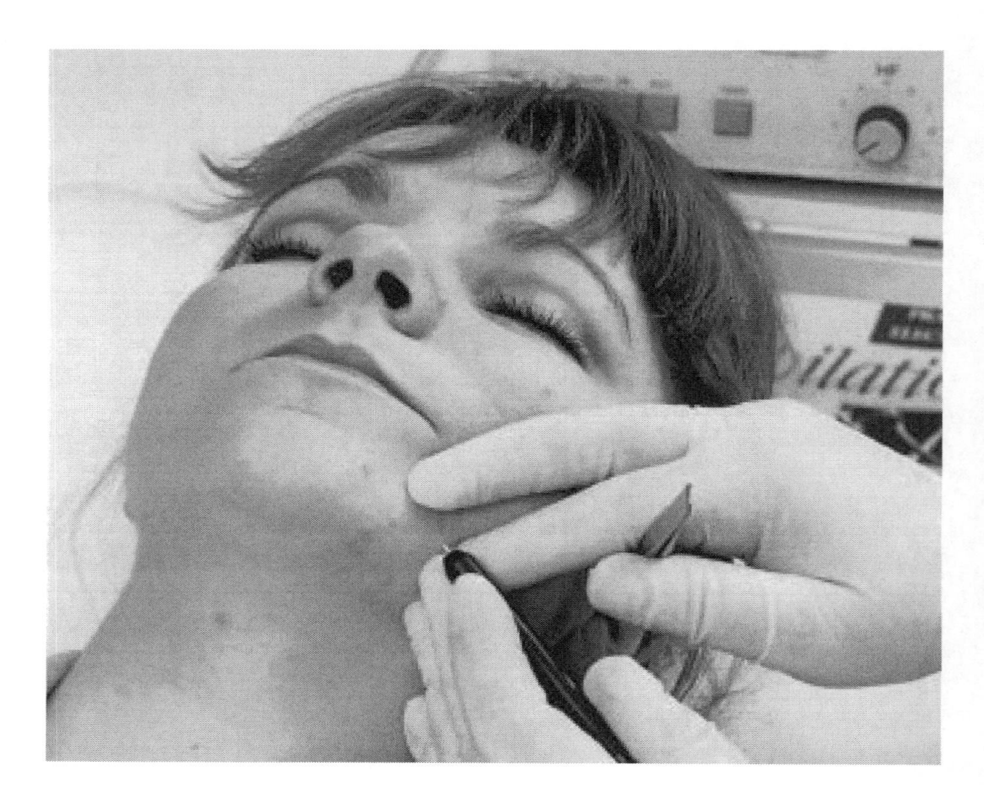

Figure 17.2j Side of jaw. Left-handed operator

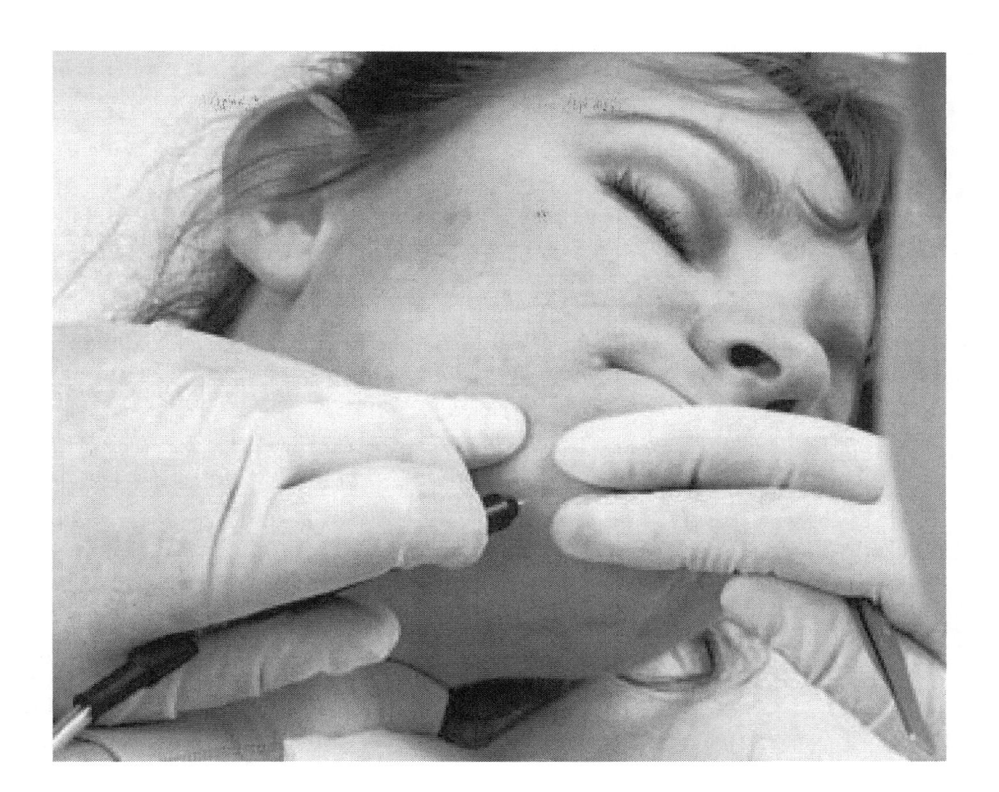

Figure 17.2k Side of jaw. Right-handed operator

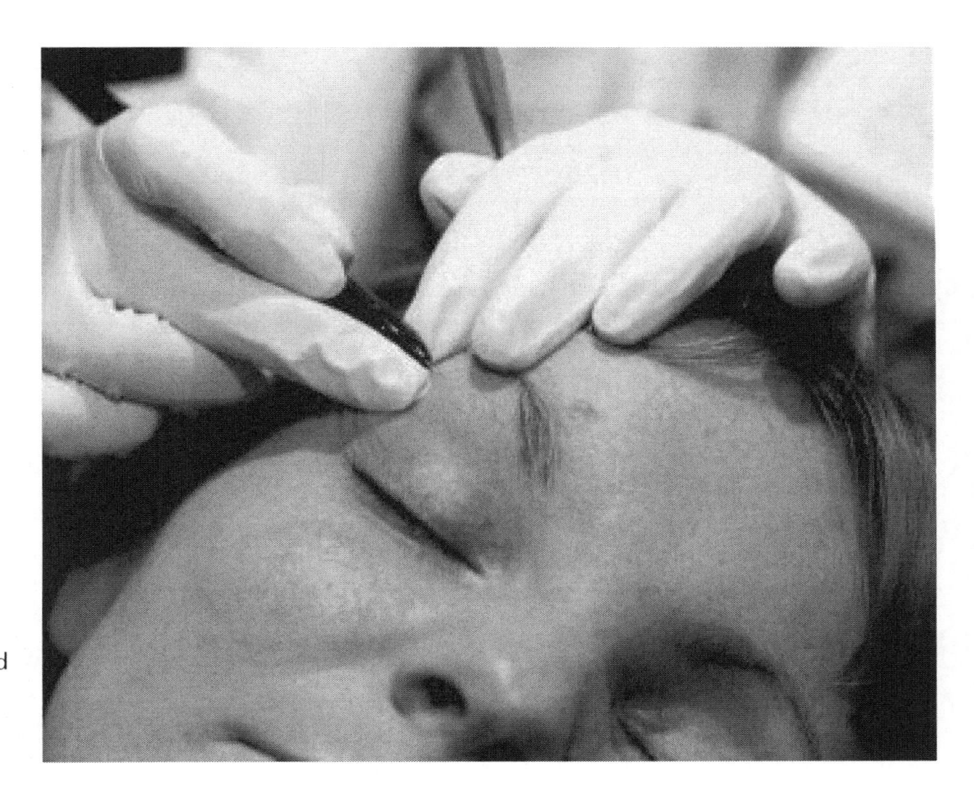

Figure 17.2l Eyebrows. Right-handed operator seated to the side of the treatment couch

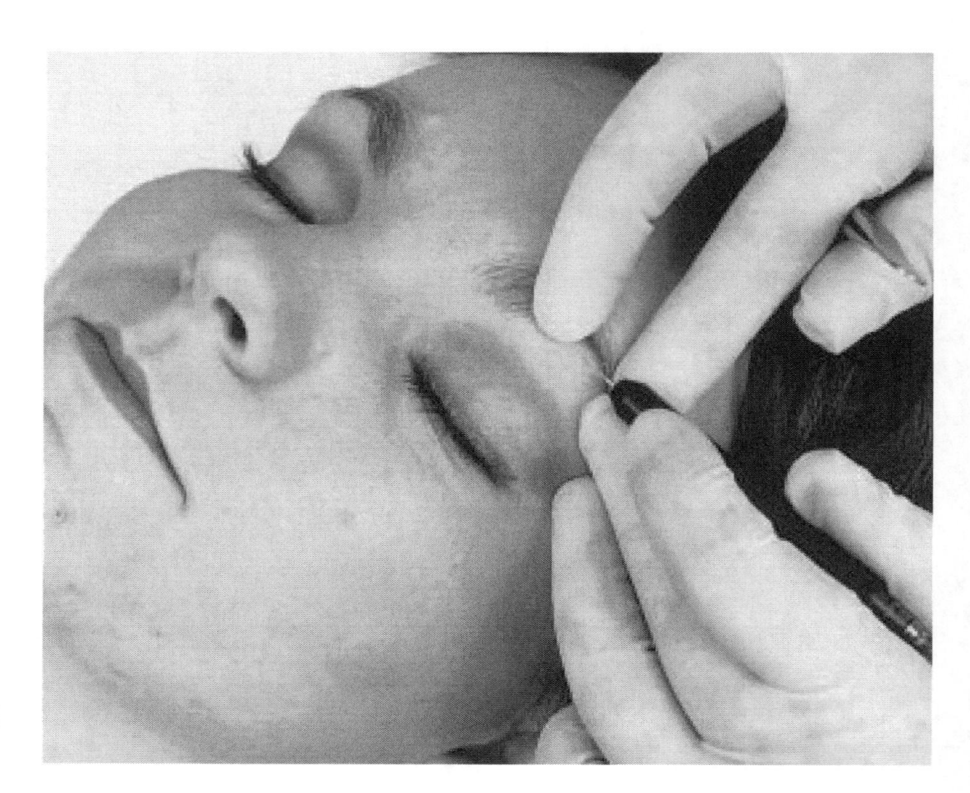

Figure 17.2m Eyebrows. Left-handed operator seated at head of couch

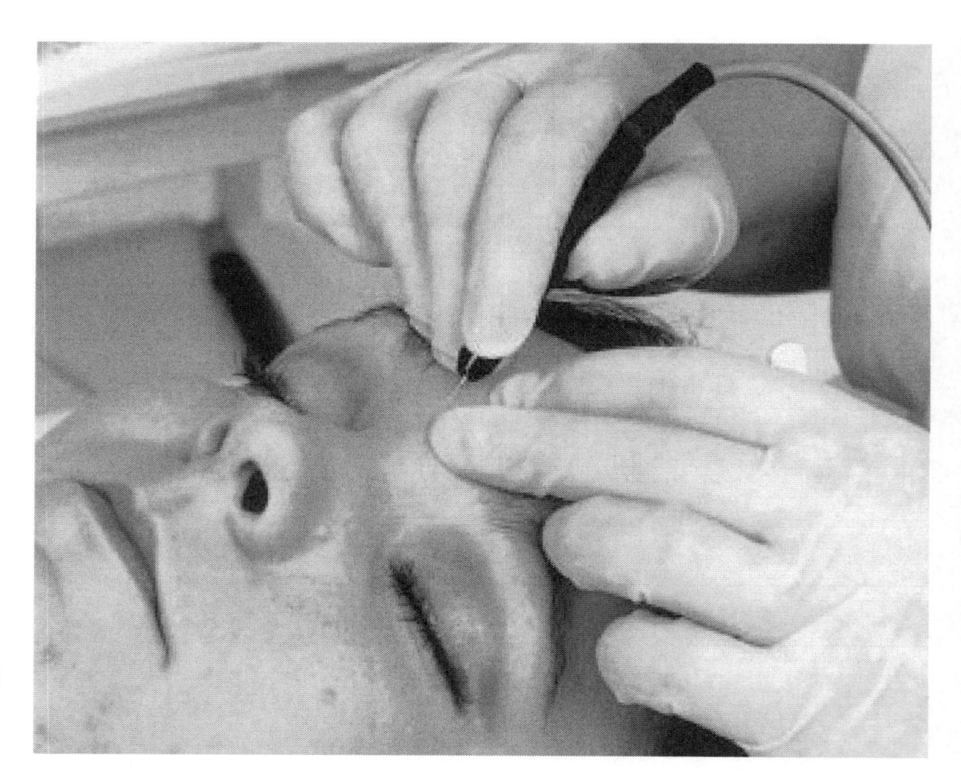

Figure 17.2n Centre eyebrows. Right-handed operator seated at head of couch

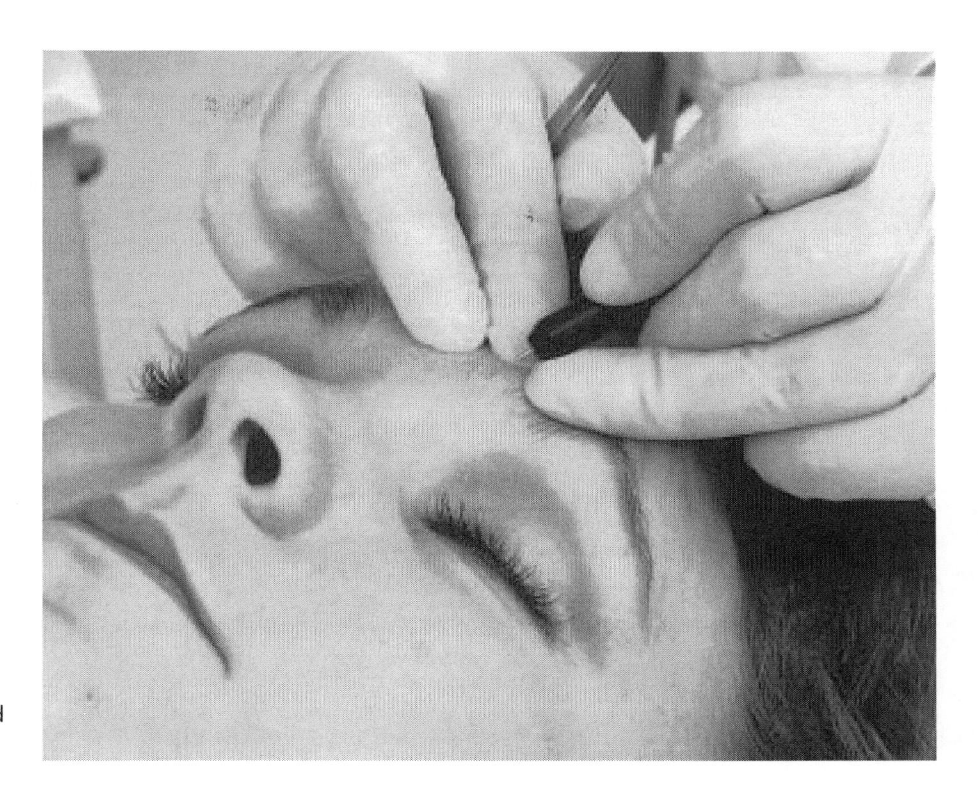

Figure 17.2o Centre eyebrows. Left-handed operator seated at head of couch

opening, which may result in bruising. There is also a risk of surface burns owing to the needle being in contact with the skin at the follicle opening.

Preparation of the area to be treated

Any make-up or lipstick in the immediate area should be removed thoroughly with a suitable cleanser. The skin to be treated should be wiped over with cotton wool and an antiseptic solution. Surgical spirit is not recommended – it has a drying effect on the skin, can be an irritant to sensitive skin and is flammable.

Technique

The stretch of the skin and the handling of the needle holder, needle and tweezers during the insertion and removal of the hair all contribute to the efficiency of the treatment.

The skin should be held firmly, but lightly. At no time should any undue weight or pressure be put on the skin during treatment. The stretch should open the follicle orifice to allow easy insertion of the needle without distorting the follicle in any way.

The hand holding the probe should rest very gently on the skin. The fingers of the opposite hand should be positioned in a manner which will not hinder the correct insertion of the needle.

Probing technique

Accurate probing is essential if treatment is to be effective. The needle should enter the follicle easily, following the direction of the hair growth. Slight resistance should be felt when the needle tip reaches the base of the follicle.

During probing, entry of the needle into the follicle should not be hindered by the fingers, as this would stretch the skin. There should not be any visible depression or puckering of the skin during insertion.

At no time should the needle be rotated in the follicle since this reduces the electrolosist's sensitivity and increases the risk of piercing the follicle wall.

Double depressions of either foot pedal or finger switch is not acceptable as normal practice. The need for double depressions indicates that the current intensity is incorrect. The application of double depressions using a finger switch increases the risk of forcing the needle through the base of the follicle on the second depression.

The finger switch should be depressed gently and smoothly. If a depression is too heavy the needle can be moved in the follicle so that it hits the follicle wall and/or pierces the base.

The angle of insertion should follow the direction of hair growth with the needle raised slightly from the skin's surface. There should be no loss of colour or depression in the skin through the needle leaning on the surface.

When the needle is resting or pressed on to the follicle wall surface burns can occur. An incorrect angle of insertion could also result in the needle tip piercing the follicle wall or entering the sebaceous glands.

Insertions which are not deep enough and do not reach the base of the follicle may give rise to surface burns due to the current being applied too near to the skin's surface. Insertions which are too deep, result in destruction of the deeper tissue, eventually leaving pit marks in the skin where the tissue has collapsed.

The needle holder should be held very lightly and without tension. When gripping the needle holder too tightly the operator's sensitivity during probing is lost.

The needle should slide smoothly into the follicle followed by application of current, withdrawal of needle and removal of hair. A correctly epilated hair should slide out of the follicle easily without signs of traction or resistance. The sequence should be smooth, rhythmic, and unhurried. Unnecessary hesitation during treatment wastes time and does not inspire the client with confidence.

Current adjustment

Sufficient current should be used to destroy the hair follicle permanently without causing the client undue pain or adverse skin reaction. The current intensity is determined by the following factors:

1. Type of hair.
2. Skin sensitivity.
3. Area being treated.
4. Client's pain threshold.
5. Weather and atmospheric temperature, for example, humidity and hot weather, bring about increased local circulation and more moisture in the skin. Therefore hairs may release easily with a lower current intensity.
6. When external temperature is cold the follicles close and nerve endings become more sensitive. This often results in increased sensitivity and a lower pain threshold.
7. Premenstrual tension and stress can also result in a lower pain threshold.

When the current is set to the correct intensity the hair will slide out of the follicle easily, complete with bulb, depending on the stage of the hair growth cycle. Dark, terminal hairs on the chin will require a higher current intensity than fine terminal, or vellus hairs on the upper lip. The client's pain threshold is much lower on the upper lip, particularly in the centre where the nerve endings are more numerous. A fine, sensitive skin will tolerate less current than a coarse or oily skin.

Incorrect adjustment of current causes a number of problems. Insufficient current results in under-treatment of the hair follicles with a higher percentage of regrowth than is acceptable. Too much current when using high frequency will result in over-treatment of skin, causing excess swelling; erythema; blanching and surface burns; too much current when using direct galvanic current will result in weeping follicles.

When the needle is too small a higher current intensity is needed in order to give enough current distribution to achieve follicle destruction.

The tweezers should be held in such a way that they do not scratch or irritate the client during probing. They should be easily accessible to the electrolysist during treatment so continuity is not lost.

Rhythm and continuity

There should be no 'stabbing' and no unnecessary selection of hairs. The treatment should be continuous without interruptions or unwarranted changes or adaptations to the routine. Check with the client to see which hairs are causing her most distress, and then treat the strongest and most noticeable first.

Pain threshold

The skin has an abundant supply of sensory nerves contained within the dermis. Many fine fibrils or nerve endings reach up into the Malpighian layer of the epidermis. These are the sensory receptors of the skin which respond to heat, cold, pain, touch and pressure. The receptors which respond to pain are known as 'free nerve endings' (see Chapter 1).

The nerve supply around the hair follicle forms a fixed network. During late anagen the follicle grows down below the network, which is an advantage to the electrolysist. When the current is applied further away from the skin's surface, i.e. at the base of an anagen follicle, the sensory nerve endings take longer to respond.

Areas such as the lip respond strongly to pain due to many nerve endings overlapping. Pain threshold varies from one area to another, and from one individual to another. It is affected by stress, fatigue, anxiety, general health and the menstrual cycle – in particular before and during the early stages of menstruation.

When treatment is concentrated on one area nerve fatigue will occur, numbing the nerve endings and lessening the sensation felt by the client. It is therefore better to work systematically in one area rather than to jump from one area to another.

The term 'pain threshold' refers to the amount of stimulation to the nerve endings that the individual can comfortably tolerate without drawing away from the source of pain. A low pain threshold means that the individual is only able to tolerate a low degree of stimulation.

Treatment organization

The organization of treatment time needs to be efficient. The client will want to feel that she is being given value for money, so time must not be wasted. At the same time she must not feel that she is being rushed in and out of the treatment room as if she were on a conveyor belt. Nor must she feel that her money is all that the electrolysist is interested in.

During the session, treat the darkest and most noticeable hairs first. Consult with the client to ascertain which hairs are causing most concern. When the hair-growth is dense it is advisable to treat every other hair. This will prevent over-treatment of the area and undue build-up of heat or sodium hydroxide. Allow sufficient healing space between the hairs. Treat all hairs of a similar size and texture. Hair texture can range from fine to coarse in any one area. When this is the case it is advisable to change the needle size during treatment. The importance of changing the needle size when necessary cannot be stressed sufficiently. The electrolysist should avoid touching the area that has just been treated.

Aftercare and home care advice

At the conclusion of the treatment a soothing, medicated cream or lotion should be applied to the area. There are a number of suitable products available, such as witch hazel based products, aloe vera and lacto calamine. Tinted aftercare preparations are also available for those clients who prefer to cover the erythema. The importance of correct aftercare must be explained to the client.

Clients should be advised not to apply make-up to the immediate area until the next day, preferably after 24 hours, at which time an aftercare cream should be applied to form a barrier between the skin and make-up. Harsh soaps and perfumed products should be avoided. Similarly the client should be advised not to go swimming where there is a risk of picking up infection and chlorine in the water will irritate the skin. Exposure to sunlight or sun beds creates heat on heat and therefore the healing rate will be slower. In addition pigmentation marks could occur. Tight jeans, clothing and tights should not be worn after treatment of the bikini line, in order to prevent friction.

Length and frequency of treatment

The length of treatment sessions is determined by the density and texture of hair growth, the client's pain threshold, the skin sensitivity and the size of the area to be treated.

The frequency of treatments will depend on the size of the area to be treated, the regrowth period, and the time the client has available.

When treatments are too close together there may be insufficient hair to treat. More importantly, the skin may not have had time to heal. When treatments are too far apart the results will be slower. The hair growth cycle should be explained to the client so that the importance of attending for treatment regularly is understood.

Client psychology can also play a part in determining frequency of treatments. When a large area, such as the legs, needs treatment a client may prefer longer treatments to clear as much hair as possible in any one session. Where an area such as the lip, chin and sides of the face need attention, the areas can be worked in rotation so that the client can have several shorter treatments more frequently, yet still avoid working in any one area before sufficient healing time has elapsed. Psychologically, the client feels better when attending more frequently.

Management of hair growth between treatments

Many clients are worried about remaining hair growth after treatment. Due to embarrassment, the majority of clients are not prepared to leave this hair alone. Therefore specific advice must be given by the electrolysist at the time of the consultation and reinforced after treatment. It is advisable to give written aftercare instructions for the client to take home with her.

The ideal is for the client to cut the hairs with scissors, close to the skin. This will not distort the follicle or affect the hair growth in any way.

When a large area is involved shaving may be the most suitable alternative. However, this can make the skin tender and will leave a shadow on the skin's surface as the hair grows.

Effect of skin type on current application

Dry skin

Dry skin may be described as lacking in sebum or lacking moisture. Skin with low moisture content is referred to as dehydrated. Needle insertion may be hindered by dead cells blocking the follicle opening. Both high-frequency and galvanic currents need moisture in order to work effectively.

Oily skin

Here there is a surface covering of sebum, which acts as an excellent insulator. This has the advantage of confining the current to the lower third of the follicle, where it is needed. The moisture content of the lower follicle is usually good. Owing to the presence of open pores, needle insertion is relatively easy. Comedones, papules and pustules may be present. The skin texture is often thicker than that of normal or dry skin. The pH balance of oily skin may be disturbed, which increases the risk of infection.

Moist skin

Moist skin has a high moisture content within both the epidermis and the dermis, therefore considerable care must be taken during current application. There is a risk of high-frequency action rising up the follicle too quickly and reaching the skin's surface before the lower follicle has been treated successfully. Blend is possibly more suitable for this skin type because a lower intensity of high frequency is used. The galvanic action will occur relatively quickly.

Sensitive skin

This is often fine in texture with telangiectasia (red capillaries). This type of skin may be dry or oily and reacts quickly to the heat produced by high-frequency. Sensitive skin often responds well to the blend technique, particularly when phoresis is applied either before and/or after treatment.

Assessing the skin's appearance during treatment

Loss of colour around the follicle during the insertion may be due to using a needle which is too large, or the operator leaning the needle on the skin.

Loss of colour during galvanic action is the result of gas and sodium hydroxide building up under the skin's surface during treatment.

Gas vapour may appear at the side of the needle during galvanic treatment, but may also result from a combination of moisture and hydrogen gas when blend is applied.

A blue or black lump indicates bruising from incorrect insertion, which pierces or damages the capillaries. Immediate pressure should be applied with dry sterile cotton wool or a cold compress.

Blanching of the skin during high-frequency treatment occurs when the current is too high or applied too close to the surface of the skin.

Weeping follicles, shortly after treatment, are an indication that too much galvanic current has been used, resulting in excessive chemical decomposition of the skin tissues.

Evaluating hair regrowth

Hair regrowth should be monitored carefully. Sometimes a client will comment that the hairs reappear within a few days of treatment. There are two reasons for this:

1. Hairs were not epilated correctly. True regrowth appears from follicles that have not received sufficient current or through inaccurate insertions that have missed the lower third of the follicle and dermal papilla.
2. Hairs are entirely different hairs to those that have previously been treated. At this stage it is advisable to go through a brief explanation of the hair growth cycle. Clients who have just started a course of epilation treatment and who have been plucking for some time are often shocked to realize how many hairs need treatment. Due to the fact that plucking/tweezing has been carried out on a regular basis, the individual is unaware of the extent of the problem.

Record cards

Record cards should contain details of treatments, for example: date of treatment; length of treatment; treatment site; needle size; current intensity; machine used; skin reaction. The electrolysist should initial entries on the card.

Accurate and well-maintained records enable the electrolysist to keep track of treatment and note progress or any changes in hair growth patterns that occur.

When equipment fails to work

There are times when the equipment fails to work properly, often at the beginning of, or during a treatment. When this happens the following procedure should be followed:

1. Check that the epilation machine is plugged into the electrical socket and switched on.
2. Check the needle holder to see if the lead/cable is attached firmly. In many instances the lead may be broken in one of three places:
   (a) where the lead connects to the needle holder
   (b) where the lead connects to the machine
   (c) a break in the middle or along the cable.
3. If there is no light showing on the machine check the fuse in the plug, and if necessary the fuse in the machine. (On receipt of new equipment it is advisable to find out where the fuse is situated.) Keep a supply of the correct fuses to hand.
4. Check that the wires to the plug are connected firmly. Sometimes with continual use one or more of the wires in the plug works loose, or the small brass screws that hold the wire in place, work loose.

Epilation Record Card Client's name

Date	Therapist	Treatment site	Method	Current intensity	Needle size	Skin reaction and comments

Figure 17.3 Record card

5 With a machine that has two needle holders, check that the correct needle holder is switched on.

6 When the magnifying lamp is not working:
   (a) check the plug as above
   (b) try changing the bulb
   (c) check the starter motor unit situated near the bulb.

For further information on safety and maintenance of electrical equipment, see Chapter 22, page 179.

Review questions

1. List the factors that influence the successful and efficient application of electrical epilation.
2. Explain the effect of the electrolysist's posture on probing technique.
3. Why is it essential for the electrolysist to work from the correct side of the treatment couch?
4. State the factors that influence the positioning of equipment.
5. What is the purpose of stretching the skin while probing?
6. Explain the disadvantage of double depressions of the button during current application.
7. What is the purpose of stretching the skin during probing?
8. How can the electrolysist prove that the depth and angle of insertion during probing is correct?
9. Explain the effects on the skin and follicle of:
   (a) shallow insertions
   (b) deep insertions.
10. What factors determine current intensity and current adjustment during treatment?
11. Explain the meaning of the term 'pain threshold'.
12. State the factors that affect the frequency and length of treatment sessions.
13. List the considerations to be taken into account when deciding which hairs to treat during a session.
14. State the advice that should be given to clients in relation to home care.
15. Describe the most effective way for clients to manage hair growth between appointments.
16. List the effects of the following on current application:
    (a) moist skin
    (b) dry skin
    (c) oily skin.
17. What effect do chemical depilatories have on the hair and skin?
18. Why are clients discouraged from plucking hairs once a course of electrical epilation has started?
19. Why is it acceptable for clients to wax or sugar the legs but not the face?
20. What is the advantage of electrical epilation over other methods of hair removal?

18 Incorrect working techniques

Incorrect working habits and poor techniques lead to an inefficient treatment and in many instances the formation of scars. In time this can result in serious loss of clients. Apart from immediate post-epilation erythema and, of course, the fact that the hair growth is either considerably reduced or no longer present, there should be no visible signs that treatment has taken place.

Blood spots, pinpricks and scabs are all undesirable after treatment, with the exception of the formation of scabs on the body or legs when either blend body technique or high-frequency flash technique have been used.

Habits to avoid

1. Stroking the hair or skin and undue hesitation between insertions leads to an inefficient treatment. Loss of rhythm and continuity could affect the client's confidence in the electrolysist's ability.
2. The hair should not be lifted or rotated around the needle before insertion. This may lead to incorrect angle of insertion and consequently inefficient removal of hairs.
3. Incorrect posture of the electrolysist is not only tiring over a period of time but also leads to faulty probing technique.
4. A tense or rigid hand and wrist during probing results in loss of operator sensitivity, and therefore the inability to sense the base of the follicle during needle insertion.
5. Holding the hair with the tweezers before and during insertion distorts the follicle, which will alter the angle of the follicle and prevent smooth insertions.
6. Double depressions, which give two bursts of current to the follicle during a single insertion, are not recommended. The second depression can easily move the needle, which can then pierce the follicle wall or the base. This can result in incorrect current distribution and consequently the formation of scars. This habit leads to excess current application and indicates incorrect current intensity.

Causes of scarring

1. Entering the follicle with high-frequency current on.
2. Coming out of the follicle with the high-frequency current on.
3. Probing through the base of the follicle with the current on. This will result in destruction of lower tissues, eventually giving rise to a depression or pit mark on the surface of the skin.
4. Probing through the side of the follicle wall with the current on. This will also lead to pit marks or depressions, owing to the current being concentrated on the wrong area.
5. Shallow probing. This allows the application of high-frequency current too close to the skin's surface. Pigmentation marks will occur, particularly in dark skin and Asian skin types.

6 Application of high-frequency current for too long, allowing too much heat to build up. Excess sodium hydroxide (lye) is produced when the galvanic current is applied to the follicle for too long.
7 Re-entering and applying current to the same follicle too often.
8 Over-treating the area. When too many hairs are treated in a concentrated area, either too much heat will be produced by high-frequency giving rise to tissue damage, or there will be too much galvanic action, which will produce 'weeping' follicles. There is also a risk of post-treatment infection.
9 When the intensity of the current is too high with high-frequency, excess heat is produced in the area. With galvanic current, too much sodium hydroxide is produced.
10 A needle that is too large stretches the follicle wall, causing bruising, and may damage surface capillaries. Current dissipates at the skin's surface where the needle touches the follicle wall, so causing a surface burn.
11 Applying treatment too often in the same area does not allow the deeper skin tissues to heal.
12 Poor-quality needles with a rough surface, blunt or broken tip, may puncture the skin or give inefficient current distribution.

The following points are not direct causes of scarring, but they might affect treatment and are therefore unacceptable.

1 When the skin is stretched incorrectly, accurate insertion is hindered or prevented.
2 Inaccurate position of hand and wrist results in the wrong angle and direction of needle entry into the follicle.
3 Incorrect position/angle of lamp casts a shadow over the area to be treated.
4 Lack of attention to aftercare may give rise to infection; chlorine/chemicals in the water when swimming may set up an adverse reaction. Sunbathing/sun beds give rise to heat on heat, so leaving the skin tender and prone to hyper-pigmentation. Applying make-up blocks the pores and fingering the area allows entry of bacteria.

Review questions

1 Name and describe five working habits that could have an adverse effect on treatment.
2 Name and describe seven causes of scarring.
3 Why is rotating the needle around the hair inadvisable?
4 Explain what happens when the hair is held with tweezers during needle insertion.
5 Why can the electrolysist's posture have an effect on electro-epilation?
6 How are the following scars caused by electro-epilation:
   (a) pit marks
   (b) brown pigmentation marks
   (c) white marks on the surface?
7 What happens to the skin when the current is applied too many times in the same treatment?
8 Explain the importance of home and aftercare following electro-epilation.

19 Hygiene and sterilization

The importance of attention to detail with regard to staff and clinic hygiene together with cleaning and sterilization practices in the clinic cannot be over-stressed. During the late 1980s mass media attention on the subject of AIDS gave rise to uneasiness and fear of the unknown among the general public and, in turn, led to clients becoming more aware of the risk of cross-infection from contaminated needles.

The electrolysist should take measures to avoid all cross-infection. Viruses that cause particular concern among clients are HIV, hepatitis B and C, which are transmitted by blood and bloodstained body fluids. Cases of infection through blood infected with HIV and hepatitis by surgery, blood transfusion, acupuncture and dental treatment have all been recorded.

The main aim is to avoid cross-infection from one client to another, and from the client to the electrolysist and vice versa.

This can be achieved easily by:

1. Carrying out thorough cleaning of treatment rooms.
2. Cleaning and sterilization of all equipment, forceps and needles.
3. Attention to personal hygiene.
4. Keeping cuts and abrasions covered.
5. Hand washing after all client contact.
6. Correct disposal of all clinic waste.

Provided that every care is taken, the risk of cross-infection during electro-epilation is minimal.

There are a number of terms with which the practising electrolysist should be familiar:

- *Bacteria:* small, one-cell micro-organisms that need a moist, warm atmosphere to survive. They also need oxygen, and they give out carbon dioxide. Some bacteria are harmless to the human body, for example, those found in the digestive tract. Others are responsible for conditions such as impetigo, food poisoning and boils.
- *Bactericide:* a chemical agent that will kill most bacteria but is not effective on viruses.
- *Virus:* minute particles that are completely inactive outside the living cells they infect. In suitable environments they are capable of reproduction and mutation (W.G. Peberdy, *Sterilization and Hygiene,* 1988). Examples of viral infections are influenza, the common cold, hepatitis A, B, C, and D and HIV.
- *Asepsis:* the absence of infection from micro-organisms.
- *Aseptic:* free from organisms capable of causing disease.

- *Sepsis:* presence of infection due to micro-organisms.
- *Sterilization:* the process used to achieve total destruction of all living organisms and spores.

HIV/AIDS

The full name for AIDS is acquired immune deficiency syndrome, which develops as a result of infection by the human immuno-deficiency virus (HIV).

The virus can be transmitted by infected blood entering the body through heterosexual or homosexual contact; entry into open wounds; contaminated hypodermic, acupuncture, or electro-epilation needles.

Cases of infection after a small percentage of blood transfusions from an infected donor in the 1980s have been recorded.

HIV-positive means the virus is present in the body and over a period of time will impair the body's defence mechanism by interfering with the immune system, which has developed from the HIV.

HIV interferes with the immune system, so reducing the body's ability to cope effectively with disease or infection such as pneumonia. It is these secondary infections which often prove fatal. A person who is HIV positive may show none, any or all of the following symptoms:

- General fatigue.
- Inability to recover fully from infections such as colds or influenza.
- Enlargement of lymph glands.
- Weight loss.
- Diarrhoea.

The HIV is a fragile virus when exposed to air and is one that is easily destroyed by the use of disinfectants. This virus is diagnosed by blood tests for anti-bodies, which may be present from between 6 and 12 weeks after infection.

Hepatitis

The hepatitis A, B, C, D, E, F and G viruses are far more resilient than HIV and are capable of existing for considerable periods of time on infected needles and hard surfaces.

The name 'hepatitis' means 'inflammation of the liver'. There are six categories of virus causing hepatitis:

1. *Hepatitis type A (HAV),* infective hepatitis, also known as short incubation hepatitis. It is spread by the faecal, oral route. The symptoms are diarrhoea and vomiting.
2. *Hepatitis type B-serum (HBV),* also known as long incubation hepatitis, affects the liver and is transmitted in the blood.
3. *HCV,* formerly known as non-A and non-B hepatitis. The incubation period is between 20 to 90 days.
4. *HDV,* also known as hepatitis delta virus. This is the only virus that is known to affect animals as well as humans.
5. *HEV,* discovered in 1990. Transmitted by drinking water infected by contaminated faeces and possibly by blood. As yet there is no vaccine available.
6. *HGV,* also known as GB virus-C. A recently discovered form of hepatitis. Transmitted by exposure to infected blood.

Hepatitis A is of short duration, with an incubation period of approximately one month. This particular virus can be contracted from contaminated food or water. The hepatitis A virus is lost from the body fairly quickly.

Hepatitis B is a far more serious condition of longer duration. The incubation period varies from between 40 to 160 days, during which time the patient is highly infectious.

The person infected with hepatitis B feels generally unwell and fatigued for a lengthy period of time, e.g. several months. This is followed by a long convalescence. Symptoms shown in the early stages include nausea, vomiting, loss of appetite and general fatigue. The whites of the eyes, together with the skin and the gums, may take on a yellow appearance. Fever may also be present in the early stages of infection. Unlike hepatitis A, the B virus remains in the body for some considerable length of time.

Contaminated needles and blood may easily transmit the hepatitis B virus. This virus can also be transmitted through needle-stick injury; contact with infected blood; through an open wound; blood transfusion; or drug users sharing needles. It has been found that the virus can remain inactive on hard surfaces for several years; therefore, high standards of hygiene in the clinic are imperative.

Vaccination against hepatitis B is possible, and indeed advisable for people working in high-risk occupations. The procedure consists of three vaccinations, the first two spaced one month apart and the third six months later. Details and information can be obtained from the doctor's surgery.

Hepatitis C is blood-borne and lies between A and B in severity. Symptoms, which are similar to those of flu, are gradual in their onset and are milder than those of hepatitis A and B. The incubation period is between 20 days and 90 days.

There is an increased risk of liver cancer in later life. A vaccine is not available for hepatitis C. Acquisition is through the use of infected syringes in drug users; where syringes are used by more than one person; post blood transfusion; and occupational exposure.

Clinic hygiene

Hygiene within the clinic is easily achieved and the procedure should be a routine matter. All equipment, hard work-surfaces and washable floors should be wiped over daily with a hospital-grade disinfectant. There are many of these available on the market, a number of which are environmentally friendly. Hand-wash basins should be cleaned regularly. Soap bars should be placed in a soap rack in-between use. Soap pump dispensers are more hygienic. Hands should be dried on disposable paper towels.

When using disinfectants for the purpose of killing viruses and bacterial spores it is essential the manufacturer's instructions for dilution percentages are strictly adhered to. Accurate timing for immersion of objects in disinfectant is important.

The disadvantage of some of the strong disinfectants is the risk of skin irritation or allergies. Some are not environmentally friendly, which cause problems when disposing of the solution.

Solutions containing chlorhexidine have a high level of antibacterial activity with low toxicity that should be used after washing with soap to disinfect the hands prior to giving treatment and preparing the skin for electro-epilation.

Once made into solution, a majority of disinfectant and antiseptic preparations have a limited shelf life and it is wise, therefore, to be guided by the manufacturer's recommendations.

Treatment couches should be covered with fresh disposable paper towels for each client. Used clinic waste, such as paper towels, cotton wool swabs, tissues and disposable gloves should be placed in a covered container lined with a plastic bag. The plastic bag should be securely tied and disposed of daily.

Surfaces contaminated by blood should be cleaned using disposable gloves and paper towels, which should then be discarded into a plastic bag.

Containers holding contaminated probes/forceps should be cleaned and sterilized daily. Instruments that have been dropped on the floor should be washed and re-sterilized prior to use.

Needles

Although it is possible to sterilize needles with an autoclave, the use of pre-sterilized, disposable needles rules out the possibility of cross-infection from one client to another, particularly as far as AIDS and hepatitis are concerned.

Disposable needles are sterilized in one of two ways:

1 Needles are packed in hospital-grade blister packs sterilized with ethylene oxide gas.
2 Individually packed needles are sterilized by gamma-irradiation – these packs have a red dot on the outside of the packet, which indicates that sterilization has taken place.

After use, the needles should be placed in a 'sharps box'.

When sterilizing needles by other methods, the following procedures should be adhered to:

- Mechanically pre-clean the needles by using swabs or cotton buds moistened with a solution of low-residue detergent or a protein-dissolving enzyme detergent and cool water.
- Needles should then be submerged in a holding container filled with a solution of low-residue detergent or a protein dissolving enzyme, then dried with tissues.
- After cleaning, the needles and instruments should be sterilized by being placed in a dry heat oven or autoclave.

Sterilization methods

Dry heat

180°C for 30 minutes
170°C for one hour
160°C for two hours

Moist heat (steam under pressure) – autoclave

Time of exposure should follow the manufacturer's guidelines:
15 minutes at 121°C (250°F) 15 psi (pounds per square inch) for unpackaged instruments/items.
30 minutes at 121°C (250°F) 15 psi (pounds per square inch) for packaged instruments/items.

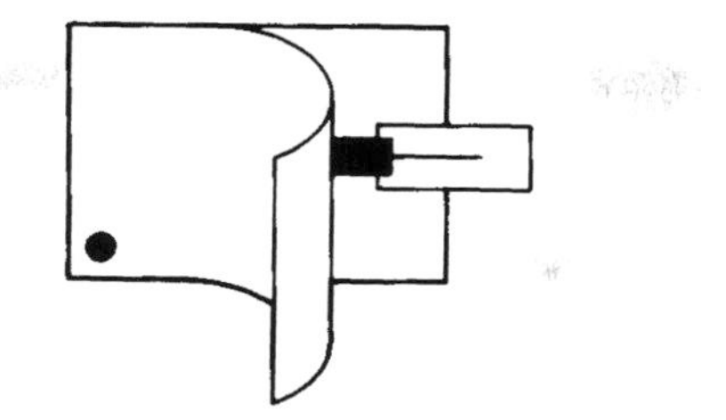

a Loosen cap of needleholder, ready to insert needle
b Peel open sealed packet; remove sterile needle with sterile forceps

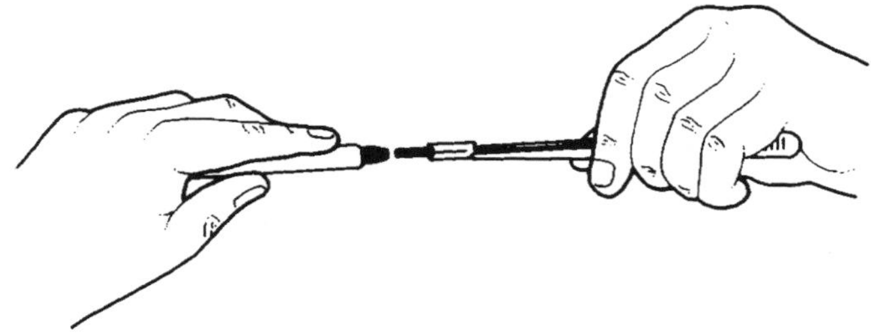

c Place needle into chuck and tighten. Gently ease off protective cap by gently twisting and pulling

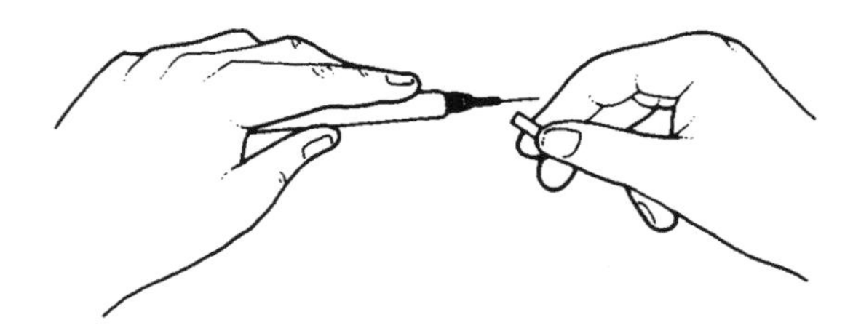

d Adjust needle by loosening cap of needleholder, using end of protective cover

Figure 19.1 Inserting sterile two-piece needle into needle holder

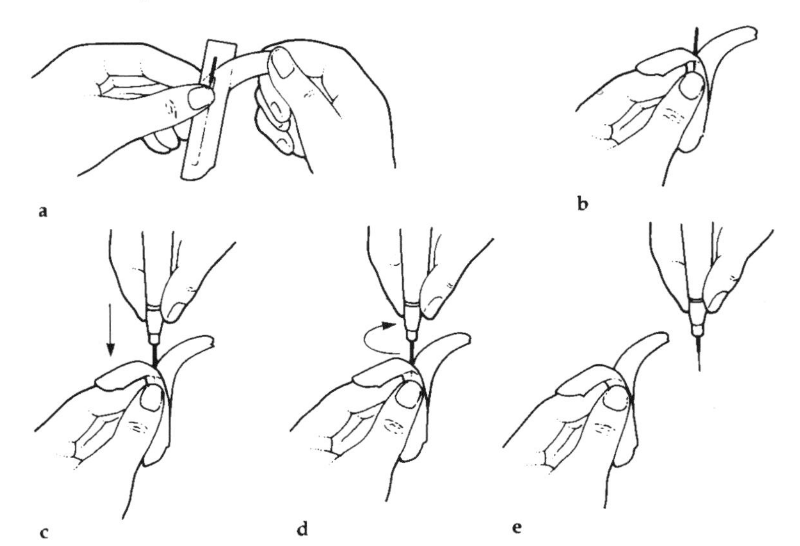

Figure 19.2 Inserting sterile one-piece needle into needle holder

The timer should be set only when the specified temperature has been reached. The heating up time is not included.

Needle-stick injury

To prevent needle-stick injury, damaged or bent needles should not be straightened or otherwise manipulated by hand, but placed in a puncture-resistant container (sharps box) that is securely sealed. The sharps box should be disposed of in the manner specified by the local health regulations or, in the USA, by the state.

Note: In the USA, the FDA have not as yet approved the use of glass-bead sterilizers. It would appear that glass-bead sterilizers do not comply with the recommendations made by the Medical Devices Agency of the Department of Health (UK) for methods of sterilization (January 1995).

Needle holders, caps, chucks and forceps

Needle holders should be wiped over with a detergent solution – germicide/disinfectant, after each treatment. Plastic or metal caps, chucks and forceps should be immersed in a disinfectant solution for at least one hour, then rinsed in water and dried with tissue or paper towel, rinsed then immersed in 70% isopropyl alcohol for at least 10 minutes. The covered container used to hold the alcohol should be emptied daily or whenever visibly contaminated.

Forceps, metal caps and chucks are all suitable for sterilization in an autoclave.

The advent of the Ballet ejector needle holder, developed by Arand Ltd, is an innovation in the field of electro-epilation. This needle holder contains an internal spring-activated mechanism which allows the epilation needle to be

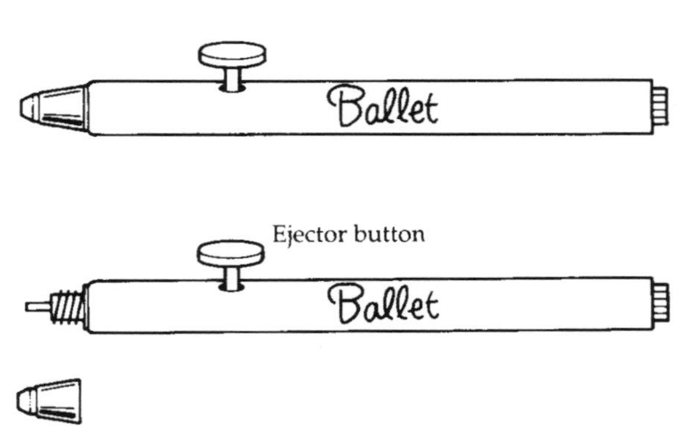

Figure 19.3 Ballet ejector needle holder

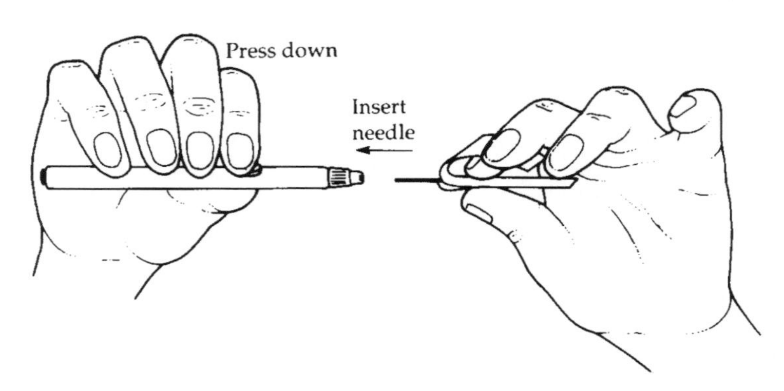

Figure 19.4 Loading ejector needle

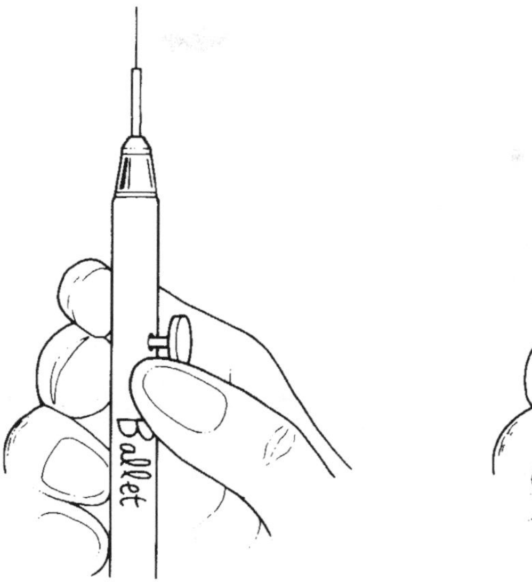

Figure 19.5 Setting the needle

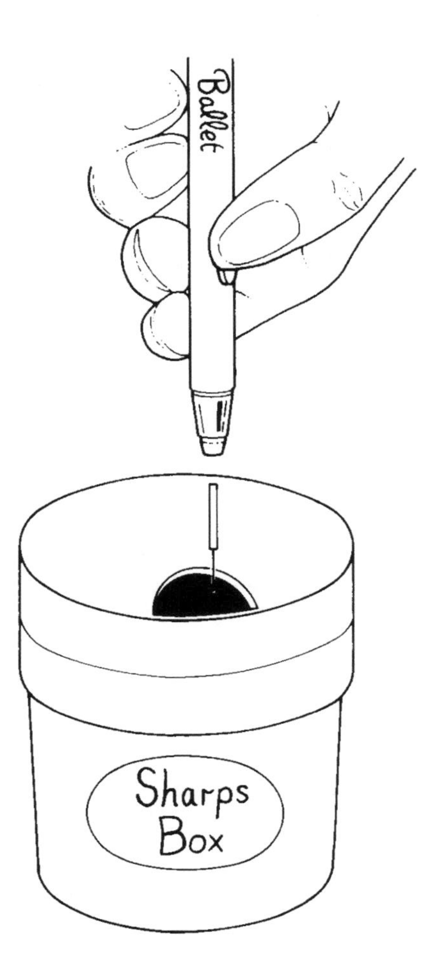

Figure 19.6 Disposing of used needle

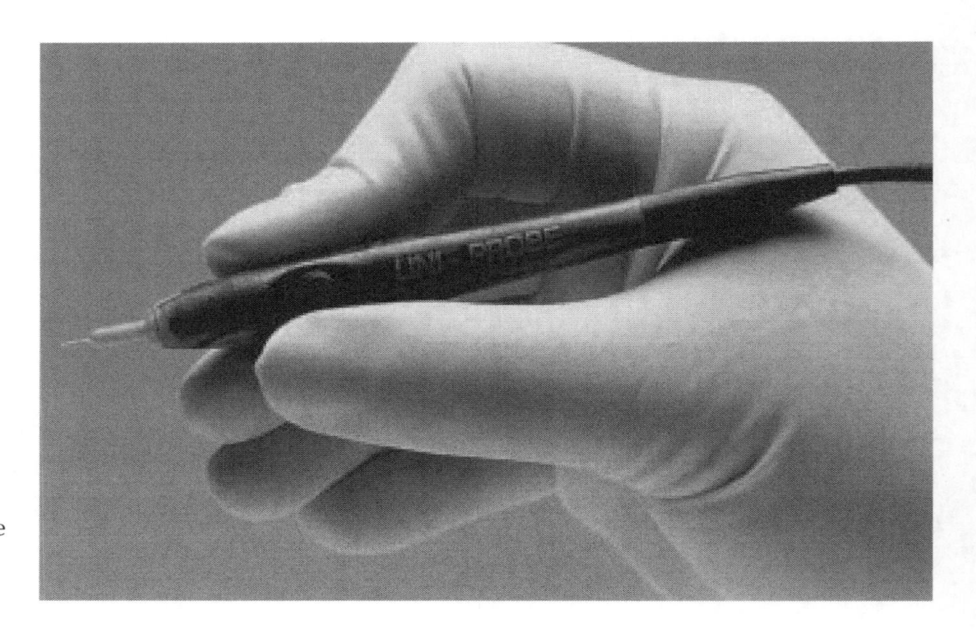

Figure 19.7 Uni-probe combined needle and cap (E.A. Ellison)

inserted and removed from the needle holder automatically when the electrolysist presses the ejector button. There is no need for the electrolysist to touch the needle with either fingers or forceps. The risk of needle-stick injury is reduced owing to the needles being ejected directly into the sharps box.

The Uni-Probe is a pre-sterilized combined disposable needle and cap, which does away with the need to wash, rinse, dry and sterilize caps.

Disposable materials

Consumables such as cotton wool pads and tissues – which may be used to protect the client's eyes against the light – and disposable gloves should be used once then discarded into a lined, covered container. The treatment couch or plinth should be covered with fresh, disposable paper towels for each client. Bin liners should be replaced daily. Sharps boxes containing contaminated needles and micro-lances should be collected by the local Department of the Environment on request or arrangements should be made with the local hospital.

Preparation of hands prior to treatment

The hands should be washed, preferably with liquid soap or bactericidal preparations, immediately before and after treatment.

All cuts and abrasions or open wounds should be covered with a waterproof dressing prior to treatment. When a number of cuts or open wounds are present, the use of disposable gloves is essential.

A fresh pair of disposable gloves should be used for each client. These should be changed when damaged or after any interruption to treatment.

Preparation of treatment site

All make-up in the area should be removed. Chlorhexidine solution or a spirit swab may be used to wipe the area prior to treatment, and the cotton wool or swab discarded immediately. The skin should then be dried with a disposable tissue prior to probing. Surgical spirit is not to be recommended because it is flammable and has a drying effect on the skin.

Sterilization methods

Sterilization is achieved in a number of ways:
Disposable needles – by ethylene oxide gas or gamma irradiation.

Metal instruments such as scissors forceps, metal caps and removal chucks – by autoclave.

Chemical disinfectants (not recommended for needles or forceps used for treatments).

Autoclaves

An autoclave is a piece of equipment that enables certain items such as forceps and scissors to be sterilized by steam at temperatures of 121°C under pressure. Modern autoclaves have thermochromic indicators which change colour when the required temperature has been reached. A stacking system is usually provided so the items can be placed at different levels in the autoclave. Temperatures range from 121°C to 134°C and correct time and heat exposure will kill all spores. However, the temperature range can unfavourably affect some materials.

Note: Information relating to the standards of sterilization of electrolysis recommended by the Centers for Disease Control was supplied by James Psnaiser, Synoptic Products, USA, and John Fantz RE, former State Examiner, USA.

Activated glutaraldehyde solutions

Glutaraldehyde is effective in destroying vegetative bacteria, spores and fungi. The effective life span of activated glutaraldehyde varies from 14 to 28 days, and therefore the manufacturer's recommendations should be followed.

Personal hygiene

The electrolysist's personal appearance and hygiene give an insight into the character of the individual. Untidy hair, bad breath, laddered tights, soiled uniform, scuffed shoes and grubby fingernails give a bad impression – one that does nothing to inspire the client with confidence. On the other hand, well-groomed hair, short, manicured nails, clean shoes, a freshly laundered uniform and fresh breath all contribute to a smart, well-cared-for appearance which will not be missed by the client.

A daily bath/shower and regular use of deodorant should be routine. Clinic uniform should be changed daily. For those electrolysists who must smoke, care should be taken to ensure tobacco odours do not cling to the breath, hair, skin, clothes and hands. Stale tobacco can be most offensive to the non-smoking client or colleague. Smoking should be confined to social hours only.

Foods that have a strong smell – such as garlic, curry or onions – are inclined to cling to the breath and should therefore be avoided prior to and during clinic hours. Teeth should be brushed regularly.

Local Government (Miscellaneous Provisions) Act 1982: Health and Safety at Work Act 1974

These acts give the local environmental health officer power to enter and inspect the premises without notice. The local authority also has the power to close down any business which does not comply with the regulations laid down in the Acts relating to standards of safety, health and hygiene. Details relating to both theses Acts can be found in Chapter 21.

(a)

(b)

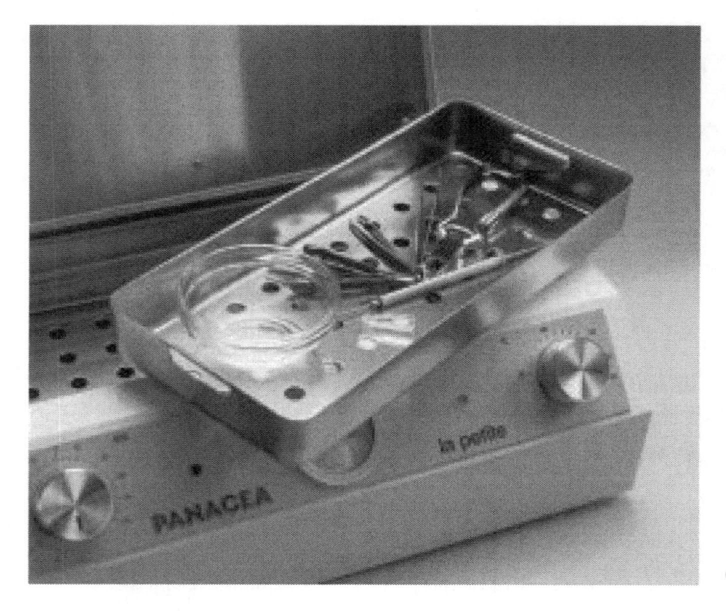

(c)

Figure 19.8 Autoclave sterilisers: *(a)* Prestige Medical; *(b)* and *(c)* Panacea

Review questions

1. Define the following terms:
   (a) asepsis
   (b) antiseptic
   (c) disinfectant
   (d) sterilization
   (e) bactericide
2. State the difference between a disinfectant and an antiseptic.
3. State the aim of hygiene and sterilization procedures in the clinic.
4. Describe the following and name two infections caused by each:
   (a) virus
   (b) bacteria
5. How does HIV affect the human body?
6. What do the initials HIV stand for?
7. What do the initials AIDS stand for?
8. List the symptoms that may be present when a person is HIV-positive.
9. Explain the difference between HIV and AIDS.
10. State the incubation periods for:
    (a) Hepatitis A
    (b) Hepatitis B
    (c) Hepatitis C
11. Explain the ways in which HIV and hepatitis B can be transmitted.
12. List the symptoms associated with hepatitis B.
13. Name two methods for sterilizing disposable needles.
14. Describe ways in which forceps can be sterilized.
15. Give the reason for covering cuts and abrasions on hands prior to giving treatment.
16. List and describe methods of sterilization.

20 First aid

It is advisable for each member of staff to undergo first aid training to the standards laid down by the St John Ambulance Brigade, the British Red Cross Association or the St Andrew's Ambulance Association (Scotland). Certificates are valid for three years, after which time a refresher course and re-examination are recommended.

The first aid manual authorized by all three associations stresses that life-saving techniques of artificial ventilation and external chest compression should not be given by a person who has not been trained in the techniques.

When giving first aid treatment the following points should be observed:

1. The situation should be assessed without endangering either the life of the casualty or the person giving first aid.
2. Identify the condition from which the casualty is suffering.
3. Administer the relevant first aid treatment.
4. Where necessary, summon a doctor or the emergency services.
5. After the casualty has been attended to, enter all details relating to the incident in a report book.

The aim of this chapter is to cover the situations that are most likely to occur in the clinic. Full details of all emergency and first aid procedures are covered in the *First Aid Manual* of the St John Ambulance Brigade, the British Red Cross Society and the St Andrew's Association, 7th edition, 1997.

First aid box

To meet the requirements of the Health and Safety at Work Act 1981, every clinic should have a first aid box which contains the following:

1. a selection of individually wrapped sterile adhesive dressings
2. sterile bandages, including triangular bandages
3. sterile eye pad
4. sterilized non-medicated dressings
5. safety pins
6. a record book for recording details of all incidents requiring first aid attention
7. cotton wool, tweezers and a pair of scissors.

The first aid box should be kept in a visible and easily accessible position.

The recovery position

The recovery position (see Figure 20.1) is used when a casualty is breathing and has a heart beat, but is unconscious. The purpose of this position is to prevent the casualty's tongue falling to the back of the throat and to keep the air passages open and free from obstructions.

1. Turn the casualty on to one side, ensuring the head is supported.
2. Place the arm at right angles to the upper body, with the elbow bent.
3. Bend the upper leg, and place at right angles to the body.
4. With the casualty's head turned to one side, tilt the head back with the jaw forward to ensure an open airway is maintained.

Figure 20.1 The recovery position

Burns and scalds

The type of burn will depend on the cause of the injury. The nature and size of the burn will determine the treatment to be given. Severe burns or burns covering a large area may give rise to shock, and hospital treatment may be necessary.

- *Dry burns* are caused by fire, lighted cigarettes, electrical equipment or friction.
- *Scalds* are caused by wet heat such as steam from a boiling kettle, very hot liquids such as water, tea or coffee and hot oil or fat.
- *Cold burns* may be caused by contact with freezing-cold metals.
- *Chemical burns* may be caused by the skin coming into contact with alkaline or acid substances. Mild chemical burns to the skin may occur when the indifferent electrode held by the client during galvanic electrolysis is not covered with damp cotton wool, viscose or similar.
- *Electrical burns* can easily be the result of too much heat being produced during an electrical current, e.g. high-frequency. This can occur when the high-frequency intensity during short-wave diathermy is too high and the underlying skin tissue or surface is burnt. Electrical burns can also be caused by lightning.
- *Radiation burns* can occur as a result of over-exposure to sunlight or snow. Rays from the sun reflect off the snow and in severe cases can cause long-term damage to either the skin or eyes.
- *Blisters* appear on the surface where the skin has been damaged by either friction or heat. They are raised swellings, which contain tissue fluid. Blisters should not be broken as there is a risk of infection occurring. In time the blister will dry up without help.

General treatment of burns

The aim is to reduce heat and pain in the area, prevent infection, and lessen the effects of shock. Medical attention or hospitalisation may be necessary in severe cases.

When possible, place the affected area under cold running water for a minimum of ten minutes. For areas that cannot be placed under cold water, bathe well with cold water. Restrictive clothing, belts or jewellery should be removed before any swelling appears, but do not try to remove anything that has become stuck to the burn. Once the pain and the heat have been reduced, cover the area with a dry, sterile dressing. Ointments, lotions and fats such as butter should not be applied to the burn.

Choking

This occurs when the air passages have become totally or partly blocked by an obstruction. This may be the result of food or drink 'going down the wrong way', and sometimes through an obstruction in the throat. The aim is to remove the obstruction and restore breathing to normal as quickly as possible.

Encourage the casualty to cough. If this fails to dislodge the obstruction, remove any false teeth from the mouth and ask the casualty to lean forward with the head in a lower position than the lungs. Apply short, sharp slaps between the shoulder blades with the heel of the hand.

Concussion

This occurs as a result of a fall, or a blow to the head or jaw, which may or may not result in loss of consciousness. The casualty may be confused and unable to remember events immediately prior to or after the incident. A careful eye should be kept on the casualty when there is a possibility that concussion has occurred, and should there be any deterioration in the condition, or should any symptoms persist, the casualty must be referred for medical attention urgently.

Carry out the procedure for loss of consciousness if necessary. On recovery from concussion, check the pulse rate and breathing. Treat for shock if applicable. Do not give liquids to drink in case the casualty needs to be removed to hospital. Head injuries can be serious and may result in long-term damage.

Diabetes

A diabetic needs to maintain a careful balance of the blood sugar level. This is usually achieved by the use of diet in mild cases, insulin injections or tablets. When insufficient food is eaten, or there is a prolonged period of time between meals, or too much insulin is taken, the concentration of sugar in the blood will drop. This can affect the brain and could result in loss of consciousness or even death.

When the blood sugar level falls, the diabetic person may feel faint and light-headed, confused or may even appear drunk. Other signs are pale skin, with profuse sweating, rapid pulse rate, shallow breathing and shaking limbs.

First aid treatment is to restore the blood sugar level as quickly as possible. This can be achieved by giving the conscious casualty sweet liquids to drink or sugar lumps to eat. If the casualty responds, give further sweetened food or liquids within a few minutes. On recovery the diabetic should seek medical advice. When consciousness is lost, the ambulance should be called immediately and the casualty transferred to hospital.

Epileptic fits

These occur when there is a temporary disruption to the normal electrical activity of the brain. Fits may be minor or major and may last from a few seconds to a few minutes.

Symptoms, which occur immediately before and during an epileptic fit, are inattentiveness, licking of the lips, incoherence and possibly loss of memory. During the fit the casualty may fall to the floor and lose consciousness.

The casualty may become rigid for a few minutes, and sometimes loses bladder control. After the fit the casualty may be a little confused and will be in need of reassurance.

The aim of first aid during an epileptic fit is to protect the casualty from injury. Clear a space in the surrounding area by moving any equipment, furniture or obstacles out of the way if possible. Should the casualty fall, place a pillow or some form of support under the head and loosen clothing around the neck.

Do not try to restrain or lift the casualty in any way and *do not* put anything in the casualty's mouth. After the fit, reassurance should be given, but drinks are not advisable until the casualty is fully alert. It will not be necessary to call an ambulance unless an injury has occurred; the casualty has taken longer than 15 minutes to regain consciousness; or has several fits in quick succession.

Fainting

This can be described as a temporary loss of consciousness brought about by a reduction in blood supply to the brain. There are a number of reasons why this may occur, including a lack of fresh air, as in a hot stuffy room; lack of food; emotional shock; and physical exhaustion.

The casualty may become pale and dizzy or may feel unsteady. The aim of first aid is to encourage the flow of blood to the brain. When the casualty feels faint, place her in a sitting position, leaning forward, with the head between the knees. Encourage deep breathing. Alternatively, when the casualty has fainted, lie her down, raise the legs and loosen restrictive clothing around the waist and chest. Ensure there is sufficient fresh air in the area.

Heat exhaustion and heat stroke

There are conditions that arise through the body becoming over-heated. Body salts and water are lost in both instances.

Heat exhaustion usually occurs after physical exercise in hot, moist conditions. Heat stroke occurs after exposure to heat or high humidity.

Symptoms of heat exhaustion are headaches, nausea and dizziness. Muscular cramps may also occur due to loss of salt. The casualty's skin may become pale, breathing may be shallow, pulse rate weak and the casualty may faint. The aim is to restore the lost body fluids as soon as possible by giving sips of cold water. When cramps are present, half a teaspoon of salt should be added to half a litre of water. The casualty should be moved to a cool place and medical aid called if necessary.

Symptoms of heat stroke are a little different, usually dizziness, nausea, headache and increased body temperature, In severe cases the casualty may lose consciousness. The aim of first aid is to reduce the body temperature as quickly as possible. This may be achieved by any of the following methods:

- moving the casualty to a cool area;
- using an electric fan to encourage the circulation of air in the immediate area;

- placing the casualty in a direct current of air;
- covering the casualty with a cold wet sheet.

A doctor should be contacted.

Insect stings

Bees and wasps cause stings, which are more painful and alarming than they are dangerous. There are some people, however, who have an allergic reaction to the poison. Stings in the mouth and throat may cause swelling, leading to asphyxia.

Signs and symptoms may be unexpected sharp pain, and swelling around the affected area. Shock may occur depending on the degree of reaction.

The sting should be removed with forceps. Pain and swelling should be relieved by the application of a cold compress, surgical spirit or a solution of bicarbonate of soda.

When the sting is in the mouth give the casualty ice to suck in order to reduce the swelling, or rinse the mouth with a solution of bicarbonate of soda and water, or cold water. Place in the recovery position if the casualty experiences breathing difficulties.

Shock

This may occur as a result of any one of the following:

- distressing news such as the death of a close relative or friend;
- an accident; the witnessing of an unpleasant situation such as a road traffic accident;
- injury to the body, e.g. electrical shock, burns or severe bleeding.

General circulation is affected, with the body directing the available blood to the vital organs, e.g. brain, heart and kidneys, leaving insufficient oxygen and blood for the remaining organs to function efficiently. Severe shock can result in death.

Signs of shock are as follows:

- the skin becomes pale, cold and moist;
- the pulse is weak;
- nausea and dizziness may be present;
- loss of consciousness in some instances.

Medical help should be called and severe cases removed to hospital. The casualty should be reassured until medical help arrives, but under no circumstances should food or drink be given which would delay the administration of anaesthetic should it be necessary.

Summoning the emergency services in the UK

In an emergency situation both time and lives may be saved when the emergency services are called quickly and efficiently. The procedure is very simple.

- Using the nearest telephone, dial 999.
- Name the emergency service required, i.e. ambulance, police, fire.
- Give the telephone number, from which you are calling, together with the exact location of the emergency.
- State the nature of the emergency and the number of casualties involved.

- Ensure that the information given is both clear and concise.
- Wait at the location of the incident for the arrival of the relevant service.

Report book

A record book should be kept on the premises. This should be used to record accurate details of any incident or accident that occurs on the premises.

The details that should be entered are:

- The date.
- The time.
- The location of the incident.
- The nature of the incident.
- First aid given.
- Action taken.

There is always the possibility that this information may be needed at a later date either for the casualty's doctor, or in the event of a possible lawsuit.

Review questions

1. Explain the purpose of first aid.
2. List the items that should be available in the first aid box.
3. Where should the first aid box be kept?
4. Describe the recovery position.
5. Explain the purpose of the recovery position.
6. Describe the first aid treatment for:
   (a) A chemical burn
   (b) An electrical burn.
7. State the first aid procedure for concussion.
8. What is the aim of first aid in relation to an epileptic fit?
9. Give the signs and symptoms of an impending faint.
10. Explain the difference between heat exhaustion and heat stroke.
11. Describe the first aid treatment for both heat exhaustion and heat stroke.
12. List the signs and symptoms of shock.
13. Describe the procedure for summoning the emergency services in the UK.
14. What is the purpose of the report book?

21 Starting and running a business

The ambition of many students at the start of training is eventually to have their own business. It is a sad fact that a high percentage of business ventures fail in the first three years. This is due to a number of reasons, but with careful planning and preparation many of the risks of failure can be minimised.

Many decide to go it alone to escape the constrictions of being an employee, or because of the shortage of good jobs available. The more positive reasons for running your own business are job satisfaction, financial rewards and independence. Those people who may be looking for shorter hours and longer holidays will be very disappointed.

Salon owners/managers are often people who have acquired professional and technical skills while working for other people/businesses and then decided to go it alone. Before long they find they are employing people in their business and need to be managers to get the best from the team. Being technical professionals, management skills very rarely come naturally. Nevertheless, success with their employees is essential for the business. Commitment is essential. Those who research their ideas thoroughly, look into all the potential problems, and take the time to produce a sound business plan are more likely to succeed.

The aim of this section is to give an insight into the planning, setting up and running of a small business. The relevant legislation cannot be covered in detail here and you will need to do some research of your own. The Bibliography lists some useful publications. A brief insight into some of the pitfalls often experienced will also be included.

Initial steps to success

1. Remember the key figure to determine the success or failure of a business is the owner. Setting up in business means being the driving force and having total commitment, self-discipline and determination.
2. Market research is essential to assess how many people are likely to require the services on offer; how many competitors are established in the area; the strengths and weaknesses of the competitors; and the scale of treatment fees charged by competitors.
3. Creating the right image for the business is important. The name chosen should convey a professional image and get the message across. It is important to check there is not another business using the same name.
4. People who make a detailed business plan are more likely to succeed. This is one of the keys to success, and it should include *realistic* cash flow and trading projections. Finance will be hard to raise without this plan. The best plans are often drawn up with professional help.

5 Finance and funding are two major hurdles. A well-prepared and laid out business plan is a valuable asset when raising finance. Your own bank is probably the first place to approach. The bank may well be able to provide all or part of the required financial support. Business premises and equipment are best financed by medium- to long-term loans, whereas day-to-day working capital is usually provided by a fluctuating overdraft facility.
6 The services of a good, professionally qualified accountant, bank manager and solicitor are essential. The services of a reputable insurance broker are also an asset. Each provides a valuable service, and the safest way of finding these professionals is by personal recommendation. The accountant will advise on finances, cash flow, VAT registration and returns, and also Inland Revenue requirements. Other areas that will need to be looked at are the insurance arrangements and the drawing up of legal documents. Legal and leasing agreements should be looked at with great care, the small print studied in detail and then checked over by experts. Every step should be taken to ensure that all legal requirements are met.
7 The location of the business is important. Thought needs to be given to the surrounding location, ease of parking, access, public transport and the number of potential clients in the area. Competition in the immediate area should also be considered. Research into any future development plans projected for nearby and surrounding areas should be undertaken.
8 Equipment should be looked at. Requirements, availability and delivery dates should be researched, as should how the equipment will be paid for, i.e. cash, hire purchase, leasing.
9 When there is the possibility that the first year's turnover will exceed the VAT registration level, it is advisable to register with Customs and Excise at the start of trading.
10 A simple accounts procedure should be set up.
11 The Inland Revenue must be informed of the proposed business. When employing staff, ensure the employment laws are met and contracts drawn up.

Business plan

When starting a business, particularly when funding is required, a well-researched and prepared business plan goes a long way to getting the business started, and raising the necessary finance. The plan should be well prepared, typed neatly and presented in a folder.

The business plan should include:

1 Name and address of the business.
2 Name of the company or person who has prepared the plan.
3 Position of the above in the proposed business, e.g. Proprietor.
4 Purpose and main activity of the business, e.g. electro-epilation.
5 Premises – location, ease of access, public transport, parking facilities. State how the premises would be set up, for example, how many treatment rooms, reception, staff room, stock room and cloakroom.
6 Benefits of the proposed business to the proprietor or partners.

7 Competition – their strengths and weaknesses.
8 Capital costs.
9 Running costs.
10 Income – sources, e.g. treatments, retail sales etc.
11 Profile of proprietor or partners, including qualifications and experience in the field.
12 References.

Locating and organizing the right premises

Having prepared the business plan and arranged finance, the next stage will be to find the right premises. Once the premises have been found, it will be necessary to contact a good solicitor to deal with all legal aspects; to keep the bank manager and accountant informed of progress; and to contact the local authority planning department and environmental health office.

Consideration should be given to the ease in which clients can reach the premises. Is public transport efficient and within easy reach? Will clients be able to park without difficulty? Look at the surrounding businesses – could they have a positive or a detrimental effect on the business, e.g. hairdressing or fashion could be beneficial, bringing clients into the area, whereas a public house or noisy record shop will have definite disadvantages!

Look at any existing businesses of a similar nature. Assess the competition – how many clients do they have? What is their price structure? Check the standard of treatments offered. Look into the size of the local population and decide whether the area can support another practice/clinic.

There are two ways of obtaining premises.

1 The property may be purchased outright by means of cash or a business mortgage. Is the property freehold or leasehold? When the property is residential, check any restrictions concerning the use of the premises for business purposes. This is particularly valid if you are intending to use part of your home.
2 When purchasing the premium on a lease for an office or retail outlet, both a solicitor and the prospective leaseholder should study the details and restrictions of the lease very carefully. Will the leaseholder or the landlord be responsible for the repairs? What is the length of the lease, e.g. five years or more? How often is the rent reviewed?

Planning department

Having found the right premises, it may be necessary to contact the planning department of the local authority to apply for change of use. The planning department will want to know whether the property is in a residential area or located within a shopping area. Where the property is residential, the planning department will look in depth at the possible effect of the proposed business on local residents. The Public Highways section will want to study the possible effect of increased traffic from clients. Will this cause a traffic hazard? Parking will be of prime consideration. Is the parking available, within the immediate location, sufficient to accommodate the number of people visiting the establishment?

Environmental health department

The environmental health department will need to be satisfied that the following government acts are being observed:

- Local Government (Miscellaneous Provisions) Act 1982
- Health and Safety at Work Act 1974 (amended 1999)
- Electricity at Work (EAW) Regulations 1989 (Northern Ireland 1991)
- Fire Precautions Act 1971
- Control of Substances Hazardous to Health Regulations (COSHH) 1999.

The Local Government (Miscellaneous Provisions) Act 1982

The Act requires that any person carrying out electrical epilation should be registered with the local authority before commencing practice. It is the operator who must be registered, not the premises. Failure to register could result in a fine of up to £200.

An inspector from the environmental health department will visit the premises and communicate with the individual before issuing a certificate of registration. The inspector will wish to know about provisions for sanitation, hygiene, sterilization of equipment and instruments used. When needles are sterilized within the clinic, the inspector will wish to know the method that is being used and assess whether it is effective. Many authorities will only accept pre-sterilized disposable needles or the use of an autoclave. The inspector will also want to know what procedure is used for the disposal of contaminated needles and waste. Storage of consumables such as bedrolls, cotton wool, tissues, antiseptic lotions etc. will be noted.

Health and Safety at Work etc. Act 1974

The purpose of the Health and Safety at Work Act 1974 is to secure the health, safety and welfare of both employees in the course of their work and self-employed persons throughout the time they devote to work.

The Act is also concerned with the protection of the individual against risks of health and safety arising out of – or in connection with – the activities of another person at work.

The employer's duty in relation to the Act is to ensure – as far as is reasonably practicable – the health, safety and welfare of his/her employees. Employers are required to draft a statement relating to arrangements for health and safety within the business.

The employee's duty in relation to the Act is to take reasonable care of the health and safety of him/herself, and of other people who may be affected by his/her acts or omissions at work. The employee has a duty to adhere to the procedures and practice of their workplace; they must never place themselves at risk through their own actions.

The Management of Health and Safety at Work Regulations 1999

The 1999 Regulations (amended from 1992) require the employer or manager to carry out *risk assessment*. This involves identifying the hazard, assessing the risk and taking measures to protect staff and clients. This assessment helps employers and employees examine every aspect of the business. Potential hazards can be identified and action taken to prevent accidents.

A hazard is anything that can cause harm, whereas a risk is the chance, great or small, that an individual will actually be harmed by the hazard.

Potential hazards can be considered under the following headings:

- physical
- biological
- mechanical

When carrying out a risk assessment include the following:

- identification of the potential hazards in all areas;
- who might be harmed and how;
- evaluation of the risks arising from the hazards;
- a decision as to whether the existing precautions are adequate or whether more should be taken.

When there are five or more employees, all significant findings should be recorded. Risk assessments should be reviewed on a regular basis and revised when necessary.

The Reporting of Injuries, Diseases and Dangerous Occurrences Regulations 1995

The Regulations require that certain accidents in the workplace must be reported on form F2508 to the relevant enforcing authority. Accidents that must be reported include so-called 'over three days injuries' – where the injury restricts the person from carrying out their full work commitments, or the person is unable to accept work at all, for more than three days. Diseases that must be reported include occupational dermatitis, legionellosis or certain musculoskeletal disorders. Further information can be obtained from the local authority or the Employment Medical Advisory Service. Accidents should be reported to the enforcing authority within 10 days.

Offices, shops and railway premises are now regulated by the Health and Safety at Work Act 1974, and are therefore subject to the general provisions of that Act.

The section covering health, safety and welfare (general provisions) relates to cleanliness, overcrowding, temperature, ventilation, lighting, sanitary and washing facilities etc. and to general provisions for first aid.

The section of the Act relating to fire precautions covers the provision of means of escape and safety in case of fire, certification of premises by the appropriate authority, fire alarms, fire prevention and provision of fire-fighting equipment.

The relevant acts can be obtained from the reference section of the local library, via the Internet or purchased direct from The Stationery Office – address and telephone number will be found in the telephone directory.

Fire regulations

Fire regulations should not be overlooked when opening a business. The local fire officer will visit the premises to inspect the provisions in the way of fire extinguishers and exits from the building.

The *Fire Precautions Act 1977* states all employees should be aware and trained in the emergency fire evacuation procedures. All members of staff should know where the nearest emergency exits are situated. These exits should be kept unlocked at all times during working hours. Staff should be aware that lifts must not be used in the event of fire or power failures;

RISK ASSESSMENT:-

Programme Area / Unit / Department:- HAIRDRESSING, BEAUTY + THERAPIES

Logical Assessment Unit:- IMAGES RECEPTION

Completed By:- DAWN EDWARDS

Date:- MAY 2003

Review Date:- DECEMBER 2003

Authorised By:- DAWN WARD

PRINCIPAL IOAN MORGAN

PERSONNEL AFFECTED – KEY:-

Staff Students	ST S	Public Contractors	P C	Young Persons	YP

RISK RATING:-

SEVERITY		LIKELIHOOD	
Fatality	3	Probable	3
Major Injury	2	Possible	2
Minor Injury	1	Unlikely	1

Activity	Personnel Affected	Hazards	Risk (Severity x Likelihood)	Existing Control Measures	Residual Risk (Severity x Likelihood)	Additional Actions
MEETING CLIENTS IN SEATED AREA	P	LARGE AMOUNT OF PEOPLE AT ANY ONE TIME	1 x 2	SIGNING IN PROCEDURES	1 x 1	
RETAIL PRODUCT AREA	P St S	CHEMICAL	1 x 2	METAL CABINET FOR STORAGE DUMMIES ON DISPLAY	1x 1	

Figure 21.1 Sample of risk assessment record courtesy of Warwickshire College

all personal belongings should be left behind; and windows and doors should be closed on leaving the premises wherever possible. Staff should also be aware of the location of fire appliances; the different types of fire extinguisher and their specific applications; and how to use them.

Both carbon dioxide and water extinguishers should be installed. Other extinguishers available are:

- Powder – for use on all types of fire.
- Halon gas – for use on electrical fires.
- Foam – for liquids.

They should be placed where they can be seen and reached easily in the event of fire.

a) A *water* extinguisher for use on paper, wood textiles, and fabric. It must not be used on burning liquid, electrical or flammable metal fires

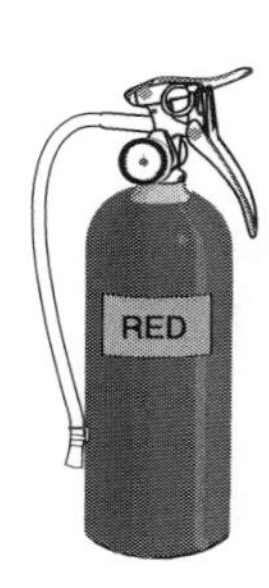

(b) A *foam* extinguisher for use on burning liquid fires. It must not be used on electrical or flammable metal fires

c) A *powder* extinguisher for use on burning liquid and electrical fires. It must not be used on flammable metal fires

(d) A *halon* extinguisher for use on burning liquid and electrical fires. It must not be used on flammable metal fires

e) A *carbon dioxide* extinguisher for use on burning liquid and electrical fires. It must not be used on flammable metal fires

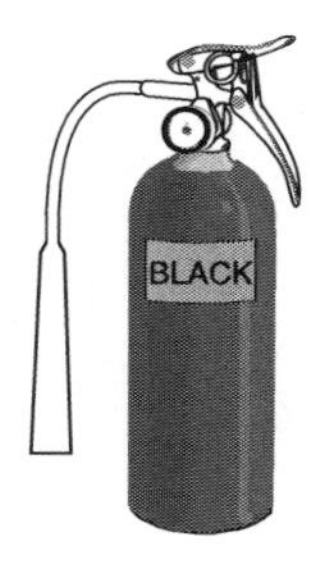

(f) A light-duty *fire* blanket for use on burning liquids and burning cloth. Heavy-duty fire blankets are available for industrial use

Figure 21.2 Types of fire extinguisher

It is advisable to have extinguishers serviced annually. A chart or placard that shows which type of fire the extinguisher is suitable for should be fixed to the wall immediately above the relevant fire extinguisher. Water extinguishers should not be used for electrical fires or near the mains supply. Carbon dioxide extinguishers can be used on all types of fire. A fire blanket is suitable for small fires.

Consumables such as paper bedrolls, tissues, cotton wool, surgical spirits etc. should be stored away from any possible risk of fire, i.e. they should not be situated under or near the mains electricity supply. Magnifying lamps should be kept away from direct sunlight. When direct sunlight is concentrated through the lens onto a flammable material a fire may result from a build up of heat.

All fire exits should be clearly marked and kept free from obstructions.

The Electricity at Work (EAW) Regulations 1989

The EAW Regulations 1989, which came into force in April 1990, were made under the Health and Safety at Work Act 1974. A 'Memorandum of Guidance' has been produced by the Health and Safety Executive (see Bibliography) and it contains a great deal of useful advice on the interpretation and practical implementation of the regulations. All electrical equipment used in the salon will be subject to inspection under the Regulations.

Control of Substances Hazardous to Health (COSHH) Regulations 1999

This law requires employers to control exposure of employees and others to substances that may be hazardous to health.

Hazardous substances include:

- Substances used directly in work activities (e.g. solvents, cleaning agents).
- Naturally occurring substances (e.g. grain dust).

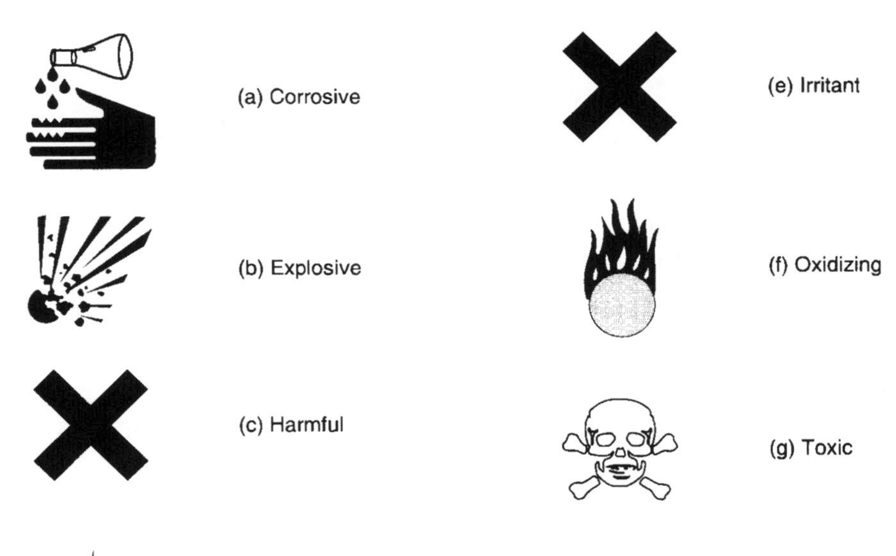

Figure 21.3 Hazardous chemical symbols

Effects of exposure to hazardous substances include:

- Dermatitis after exposure to an irritant.
- Allergies such as asthma after exposure to certain fumes.
- Infection from bacteria and viruses, e.g. impetigo, hepatitis and AIDS.

There are several steps that employers must take in order to comply with COSHH regulations. These are:

- Assess the risks.
- Decide what precautions are needed.
- Prevent or adequately control exposure.
- Ensure control measures are used and maintained properly and that safety procedures are followed.
- Monitor the exposure of employees to hazardous substances if necessary.
- Ensure employees are properly informed, trained and supervised.

Additional basic legislation

The Sale of Goods Act 1979

This Act states that goods offered for sale must be fit for the purpose for which they are intended.

Supply of Goods and Services Act

The purpose of this Act is to extend the protection given to consumers under the Sale of Goods Act to include services, goods on hire and goods given in exchange for gift vouchers.

Trade Description Act 1968 and 1972

This Act states that the retailer must not give misleading information to the consumer about goods and services, or make dishonest comparisons relating to prices. Descriptions of both goods and services must be accurate.

Data Protection Act 1984

Computers are fast becoming an essential piece of office equipment. It is now a legal requirement that when storing personal information (records) relating to clients the user should register with the data protection registrar. Clients are now entitled to ask for copies of any information which has been stored and which concerns them. It is therefore important to appreciate that only information of a professional nature is recorded, together with appointments etc. It is not wise to store any personal observations or comments. The application forms DPR1 or DPR4 (intended for use by small businesses) are available from:

The Data Protection Registrar
Springfield House
Water Lane
Wilmslow
Cheshire SK9 5AF
Telephone 01625 535 7777

Holding of personal data on a computer by an unregistered person is a criminal offence.

The role of computers in the clinic

A computer can be an invaluable piece of equipment when chosen wisely and used correctly. A computer is ideal for keeping the business accounts, producing cash flow and profit forecasts, stock control, clients' records, treatment details and for word processing.

The computer is also a valuable marketing tool via the Internet and websites. For more information on this, see Chapter 23.

Value Added Tax

VAT, or Value Added Tax, is levied on most business transactions that take place within the European Community (EC), the UK and the Isle of Man. This tax is called by different names and is of varying amounts in different countries. Each country is responsible for its own rules and regulations relating to this tax. Her Majesty's Customs and Excise are responsible for the collection and administration of VAT in the UK.

It is a legal requirement for all businesses within the UK and Isle of Man to register with the VAT office when trading turnover is likely to reach or exceed the threshold limit. The VAT threshold is set by the government and reassessed annually. The threshold limit at April 2001 was set at £54,000 annual turnover. VAT is charged on taxable turnover, which is the value, not just the profit, of all taxable supplies made in the UK or Isle of Man. For example, where the rate of VAT is levied at 17.5% on sales and treatments the breakdown would be as follows:

Treatment fee	£10.00
Add VAT at 17.5%	£1.75
Treatment fee to client	£11.75

Usually, price lists for treatment fees are shown with the VAT included in the price.

The breakdown to the business would be as follows:

Treatment fee to client	£11.75
Less VAT at 17.5%	£1.75
Portion due to business	£10.00

The above is a simple example and applies to all sales and purchases made by the business. The business is able to offset VAT paid on purchases against VAT received from sales and treatment fees. The balance is then paid over to Customs and Excise where the VAT on receipts exceeds VAT on purchases. Certain services and goods are exempt from VAT, such as general rates.

Failure to register the business for VAT in time may result in the levying of a financial penalty. VAT returns should be completed and presented to the office at specified times, with financial penalties being imposed for late returns.

Full details relating to VAT can be obtained through the Internet or through the local VAT office.

Inland Revenue

The Inland Revenue will want to know the name of your accountant, the type of business you are running and whether you are a sole trader or in partnership.

At the end of each trading year, a copy of the business accounts should be sent to the tax inspector. Tax will be based on the business income, less allowable business expenses. The accounts will need to satisfy the inspector that they show a true reflection of the business.

When taking on employees it may be necessary to deduct income tax and national insurance contributions from their wages/salary, as well as pay the employer's share of the national insurance contribution.

Employees normally pay income tax on the PAYE (pay as you earn) basis. The tax office issues an employee with a tax code. Using this code, together with the tax tables supplied by the Inland Revenue, the employer deducts income tax from the employee's salary or wages each payday. This is then paid over to the Inland Revenue on a monthly or quarterly basis.

The tax year runs from 6 April to 5 April of the following year.

At the end of each tax year the employer will issue the employee with a P60 form. This form give full details for the wages/salary paid to the employee during the financial year, the income tax paid and the national insurance contributions paid. The employee must keep this form in a safe place. The P60 cannot be replaced if it is lost.

It is essential to keep full and accurate records of business accounts for your own use, as well as for use by your accountant and the Inland Revenue. Businesses registered for VAT may also be requested to produce their accounts for inspection.

Self-assessment

Anyone who is self-employed or receives a **tax return form** from the Inland Revenue is required to complete this eight-page form, listing details of all taxable income and gains received during the year and claim any allowances. The Inland Revenue will work out the amount of tax to be paid providing the tax return form reaches the Inland Revenue by 30 September following the end of the tax year on 5 April. To avoid fines or penalties, individuals who decide to work out their own tax bill must send the completed the tax return to the Inland Revenue before the following 31 January.

National Insurance Contributions

All employed and self-employed persons pay national insurance contributions unless exemption has been granted.

The employer will make deductions direct from salary, at the same time that income tax is deducted. The amount deducted from income will be earnings-related. Both the employer and the employee will make contributions.

The employee will be entitled to claim sickness and invalidity benefits, unemployment benefit, maternity benefit and widow's benefit when sufficient contributions have been made. A portion of the contributions goes towards the retirement pension.

The self-employed, pay class 2 national insurance contributions. The most straightforward method of payment is by direct debit through the bank. These should be paid weekly or monthly, at a flat rate regardless of earnings. At the end of the tax year a further earning-related class 4 contribution may be required, based on a percentage of the taxable profit of the business.

Benefits available to the self-employed are sickness and invalidity benefit, and basic maternity allowance. A contribution will be made to the basic retirement pension.

Full information relating to national insurance contributions can be obtained from the local employment office and the Department of Social

Security. The telephone numbers will be in the local directory. Information may also be obtained through the Internet.

Costing of treatments

This is an area that is very often neglected, without sufficient thought being given to the long-term future of the business, or income for the self-employed. The temptation by many self-employed therapists and new businesses is to undercut the local competition. It is very difficult to increase prices in a short period of time when the initial pricing structure was incorrect.

The treatment fee should reflect the overheads, costs and an allowance for a profit margin. Psychology also plays a part in the pricing of treatments from the client's point of view. When the price is too low, the client may not have confidence in the electrolysist's ability to give an effective treatment. On the other hand, when the price is too high and does not reflect value for money, the client will look elsewhere.

The trading costs for a 12-month period should be looked at in detail. These include rent, rates, electricity, heating, lighting, replacement and maintenance of equipment: insurances; consumables such as needles, cotton wool, paper bed rolls, tissues etc.; wages, telephone, advertising costs, stationery and printing costs, accountancy fees, bank charges and licence fees.

The number of trading hours available should be considered and allowances made for bank holidays and traditionally quiet periods such as January, February and perhaps August. An assessment of the overheads and running costs can be made from the above, and a realistic treatment fee arrived at.

Insurance

Insurance is an essential consideration when setting up in business. The hope is that a claim will not be necessary, yet realistically no business or practising electrolysist can afford to be without adequate cover. Who knows what may happen in the future – possibly through no fault of your own?

Insurance policies required by law include car insurance, employer's liability when employing staff and public liability.

There are also a number of other policies that should be considered when setting up in practice, either as an employer or as a self-employed electrolysist. Some are necessary from the start of business, others can be added at a later date as and when funds permit.

Treatment risk protects the electrolysist against legal liability for accidental injury or damage to a client arising out of treatment, e.g. scars from incorrect electrical epilation, injury to the eye, or permanent damage that may result in psychological distress. The onus is on the client to prove that the injury or damage suffered is due to the fault of the electrolysist.

Employers' liability is a legal requirement when staff are employed. This policy covers the employers for legal liability in the event of injury to an employee arising out of their employment. A valid Employers' Liability Certificate must be displayed on the premises. Certificates must be kept for at least 40 (Yes, forty!) years.

Public liability covers the electrolysist or salon owner for the accidental loss or damage to property and for accidental injury to a third party (e.g.

a client – not your own property) arising in connection with the business. For example, a client may trip on a defective carpet during a visit to the premises and as a result sue for damages.

Building and contents should be covered against fire, flooding, storm damage, gales, burglary, theft and other specified perils. Accidental damage is standard in the industry. In comparing the cost of policies it is sensible to check that the excesses are the same. If you own the building you should check that subsidence cover is included. The car must also be insured for business use.

Car insurance for business purposes is essential for mobile electrolysists. Cover should be extended to include equipment in the car on an All Risks basis, also stock, personal possessions or money.

Business interruption: the policy covers interruption to business following a disaster such as fire, flood, storm etc. and extends to include denial of access and failure of public utilities. The standard indemnity period is 12 months but this can be increased to 24 or 36 months. Cover extends to include:

- Loss resulting from murder, suicide or disease.
- Damage in the vicinity of the clinic preventing access.
- Damage at suppliers' premises.
- Damage at public utilities (gas, water, electricity, telecommunications).

Permanent health insurance provides long-term protection for loss of income arising through illness, injury or accident. When self-employed, illness or permanent disability can result in serious financial hardship. Commitments such as mortgage, council tax, electricity, gas, telephone and food still have to be met.

Statutory sick pay, incapacity to work and invalidity benefits provided by the government are designed only to protect against poverty and do not provide sufficient income to maintain standards of living. The premium for permanent health insurance will vary, depending on the amount of benefit required, whether there is a deferment period of four weeks or more before benefit may be claimed, and the general health of the person insured.

Personal accident insurance and sickness pays specific benefits (e.g. death, loss of eye or limbs) owing to injury following an accident, provided the cause is not connected with high-risk activities such as winter sports. This cover does not include illness of any type.

Hand disablement covers the hands below the wrist and includes illness and accident, e.g. broken finger or contracted dermatitis.

It is advisable to contact a registered insurance broker for advice concerning the relevant policies. Registered insurance brokers are able to recommend the most suitable policy for your individual needs.

Health and Safety legislation: although, at present, there is no specific statement with regard to the requirements for Health and Safety legislation on insurance policies, there is a general condition on every policy that all laws, local authority requirements etc. must be complied with. For example, salon /clinic owners need to have a written Health and Safety Policy, therefore, in the event of a claim, it would not be unreasonable for insurers to ask to see it.

Receptionist

The receptionist plays a key role in the smooth running of the salon. It is the receptionist who provides the initial point of contact and is usually the first person the client sees or talks to when entering the salon. The receptionist should be welcoming, professional and friendly without being over familiar. The client must be made to feel welcome and at ease.

It is an advantage for the receptionist to be familiar with epilation treatments and be able to explain the procedure when a prospective client makes a general enquiry. The receptionist should encourage the person concerned to attend for an initial consultation with an electrolysist.

The receptionist should have the ability to deal with all manner of people without becoming flustered, harassed, argumentative or intolerant.

Telephone enquiries and bookings

The telephone often forms the first point of contact between the salon and the prospective client. Clients may be nervous or know very little about electro-epilation when making the initial enquiry.

During this telephone conversation the receptionist or electrolysist should give a brief explanation of the procedure and purpose of electro-epilation. Information regarding the value of a clinic consultation with the electrolysist should be given.

The name and address of the enquirer can be taken so that details of electro-epilation may be sent through the post if desired. The advantage of forwarding details of the clinic and treatment procedure is that it enables the recipient to read the relevant information prior to the initial consultation.

The person taking the initial telephone call should have a warm, pleasant telephone manner together with a clear voice. The caller should not be kept waiting on an open line. It is irritating to the caller and leads to dissatisfaction. When booking appointments over the telephone state the date, time and day of the appointment clearly, then confirm by repeating these details.

At times it may be necessary to use the facility of an answerphone. In this situation the outgoing message should be clear and any messages received should be dealt with at the earliest opportunity.

Review questions

1 What is a business plan?
2 Explain the purpose of a business plan.
3 List nine initial steps to success that should be considered when setting up a business.
4 List the main points of the following legislation:
   (a) Local Government (Miscellaneous Provisions) Act 1982
   (b) Electricity at Work Regulations 1989
   (c) Health and Safety at Work Act 1974
   (d) COSHH (Control of Substances Hazardous to Health) 1999.
5 What is the purpose of the Health and Safety at Work Act 1974?
6 State the requirements of the Management of Health and Safety at Work Regulations 1999.
7 Describe the procedure for Risk Assessment in the workplace.
8 Give a brief description of the Fire Precautions Act 1977.
9 State which type of extinguisher is used for each of the following fires:
   (a) Paper.
   (b) Electrical.

(c) Liquid.
(d) Wood.

10 What is covered by the COSHH Regulations 1999?
11 Define VAT.
12 When should a business register for VAT?
13 Define PAYE.
14 Define National Insurance contributions.
15 Explain the term 'self-assessment' in relation to the Inland Revenue.
16 List the points that should be taken into consideration when setting treatment fees.
17 State the purpose of the following insurance policies:
(a) Treatment risk
(b) Public liability
(c) Employers' liability
(d) Business interruption.
18 Explain the role of the receptionist in the smooth running of the salon.

22 Epilation equipment

Choosing and buying the right equipment at the start can save hours of frustration and loss of business because of poor-quality equipment breaking down, often at the most inconvenient moment. Buying the cheapest equipment because it looks like a bargain often turns out to be false economy.

The needs of the mobile electrolysist will be different from those of the electrolysist who works in a salon. The mobile electrolysist will need:

- Lightweight equipment that is easy to transport.
- Equipment that can be set up quickly.
- Sturdy equipment that is not easily damaged in transit.

Choosing suppliers

Suppliers of both equipment and consumables that are reliable and efficient are an asset to any well-run business.

Equipment should be chosen with care. The following questions should be answered:

- Is the equipment well made, safe and reliable?
- Does it conform to British and European Safety standards?

Then enquire into delivery dates, service and repair facilities. Check the supplier offers a back-up service should it be necessary. Some computerized equipment and more sophisticated machines are not easy to use at first glance, so make sure training in the use of the equipment is provided.

When choosing suppliers for consumables look to see if there is a good cash and carry in the area. If there is, are the premises easy to get to? How do the trading hours compare with those of the proposed business? If you are intending to use mail order, find out how reliable or efficient the service is. Goods supplied through mail order may be slightly more expensive to take into account the handling costs involved. Then check the quality of the goods and that they are up to the standard required. The final question must be: are the goods readily available when they are needed? Some suppliers offer a delivery service.

A good working relationship with the right supplier can be of great value to the business.

Choosing equipment

When buying equipment a number of points need to be considered before making a final decision:

- Is the equipment going to be used in the salon and therefore remain in one position without being constantly moved around? Is it intended for use in a mobile practice where the equipment will be moved constantly? (When used in a mobile practice the equipment needs to be light and robust.)
- Does the equipment conform to European Regulations (EU countries only)?

- Is the equipment well made, *reliable* and durable?
- Enquire into delivery dates, training if required, delivery dates and repair facilities.
- Does the supplier offer a back-up service?
- Is the quality of the equipment reflected in the price?

Maintenance of equipment

The Electricity at Work Regulations 1989 state that a qualified electrician should check all pieces of electrical equipment within the workplace regularly. The electrician will then label the equipment, stating the date of inspection.

The risk of injury from electricity is strongly linked to where and how it is used. The risks are reduced considerably when the following procedures take place on a regular basis:

- Choose equipment that is suitable for its working environment.
- Ensure equipment is safe when it is supplied and then maintain it in a safe condition.
- Some types of equipment are double insulated. These are often marked with a double square.
- Check plugs, leads, and wires to ensure there are no loose connections.
- Equipment should be cleaned daily.
- Provide an accessible and clearly identified switch near to each fixed machine to cut off power supply in an emergency.
- All equipment should be situated near to an easily accessible electrical socket-outlet, so it can be easily disconnected in an emergency.
- The ends of flexible cables should always have the outer sheath of the cable firmly clamped to stop the wires from being pulled out of the terminals.
- Replace damaged sections of cables completely.
- Protect light bulbs around the magnifying lamps and other equipment that could easily be damaged in use as there is a risk of electric shock if they are broken.
- The supply leads have only two wires – live (brown) and neutral (blue). Make sure they are properly connected if the plug is not a moulded-on type.
- Epilation machines should be kept on a suitable trolley during use.
- Castors on trolleys and stools should be looked at to ensure they are working efficiently and are not broken or loose.
- Electrical leads and wires should not be allowed to trail on the floor.
- Always check the machine is working correctly before commencing treatments.
- Equipment should not be used on wet surfaces or where water could give rise to an electric shock.

Check that:

- Suspect or faulty equipment is taken out of use and labelled 'DO NOT USE'. It should then be examined by a competent person.
- Where possible, tools and power socket-outlets are switched off before plugging in or unplugging.
- Equipment is switched off and/or unplugged before cleaning or making adjustments.

Epilation equipment for the treatment room

Treatment couch

The treatment couch needs to be comfortable for the client, well padded and preferably have an adjustable lifting head. From the electrolysist's point of view, the couch needs to be accessible, allowing the electrolysist to work comfortably with her knees under the couch. An electric or hydraulic couch that makes it possible to adjust the height is a real asset to the operator who spends much of her working day giving electro-epilation treatments.

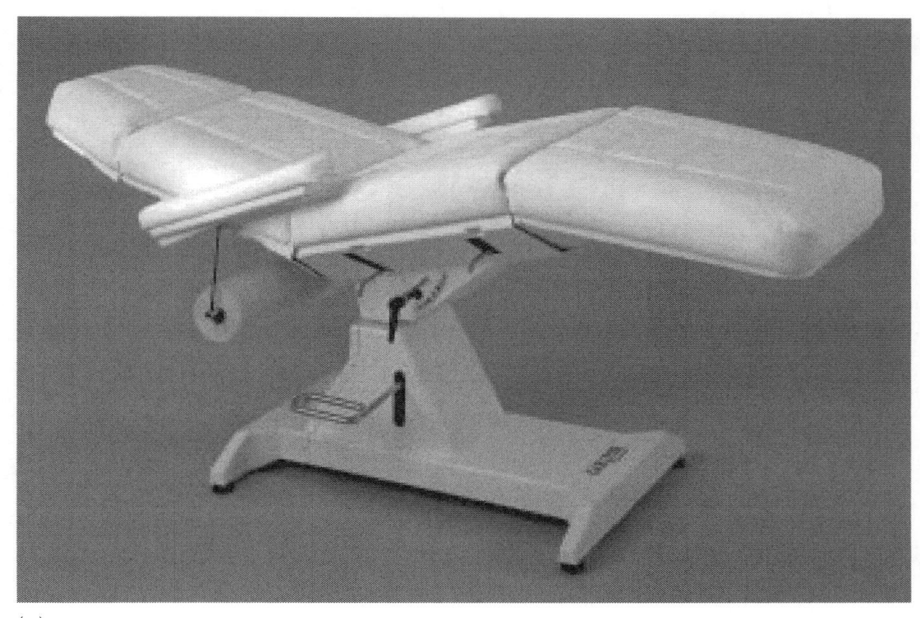

(*a*)

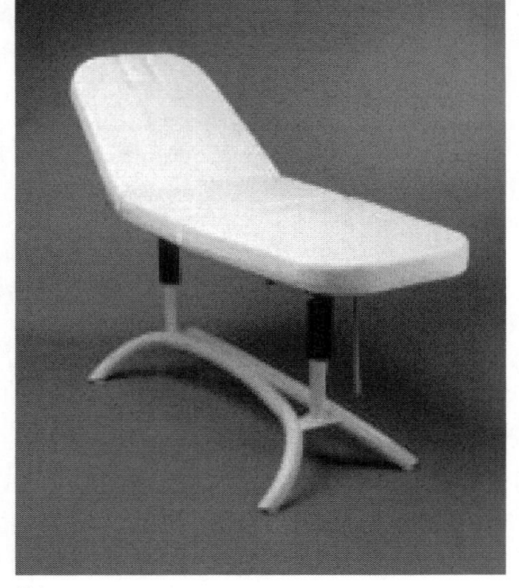

(*b*)

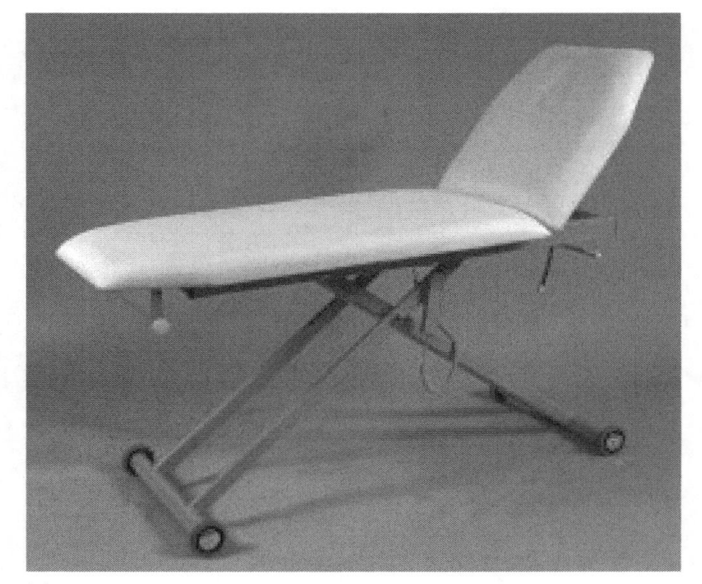

(*c*)

Figure 22.1 Treatment couches: (*a*) Carlton Professional; (*b*) Astra (House of Famuir); (*c*) E.A. Ellison

Trolley

There is a wide selection of trolleys available. The trolley should be chosen to suit the size of the treatment room and the size of the electro-epilation unit that it is going to hold. Some trolleys have an electric side panel, some have lower drawers and others have a facility for attaching the magnifying lamp. Castors, to allow easy movement, are an advantage for all trolleys.

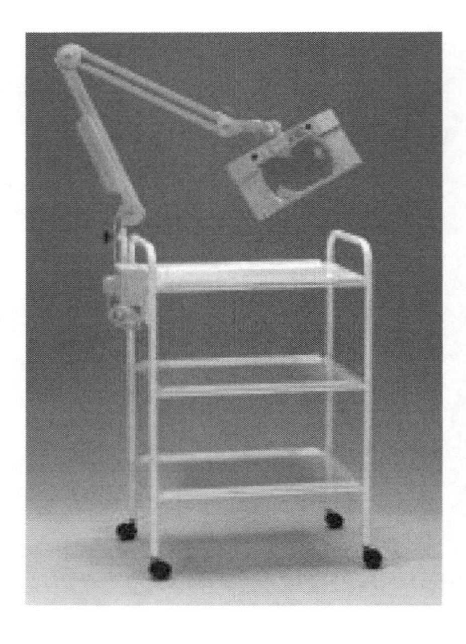

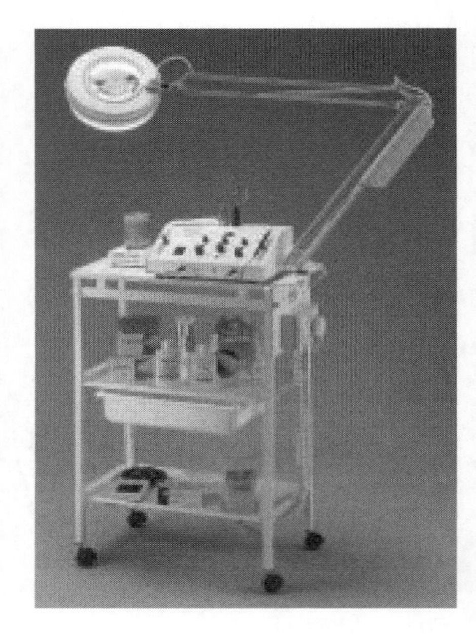

Figure 22.2 *(left)* Trolley and magnifying lamp (E.A. Ellison; *(right)* trolley and equipment (Carlton Professional)

Magnifying lamp

These can often prove to have minds of their own during treatment. Just as the electrolysist has the lamp in the right position and is about to start treatment, the lamp drifts off – as if by magic! The lamp needs to provide good light and magnification of the area being treated. The head of the lamp should remain in position, without causing distortion. The whole unit should be well balanced so that it does not overbalance during treatment.

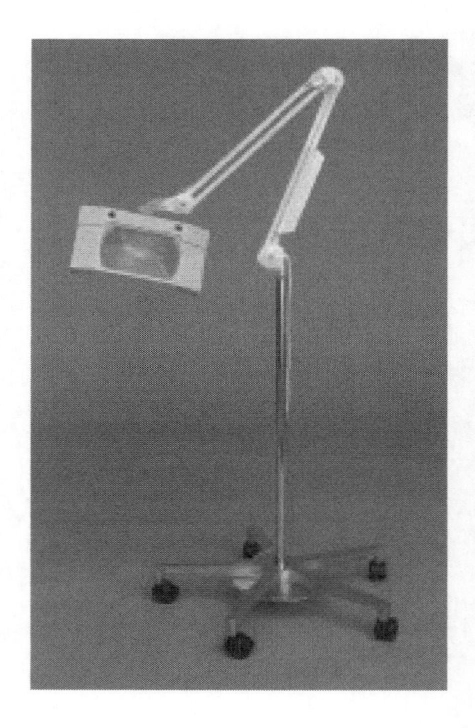

Figure 22.3 Magnifying lamp E.A. Ellison)

Operator's chair/stool

A suitable chair or stool makes all the difference to efficiency of treatment and comfort of the electrolysist during treatment. The ideal is a chair that has castors to enable to electrolysist to move around the treatment couch with ease. The height should be adjustable, preferably with a gas-lift action – which makes it possible for the height to be raised or lowered in seconds. The addition of a back-rest gives the electrolysist extra back support during treatment.

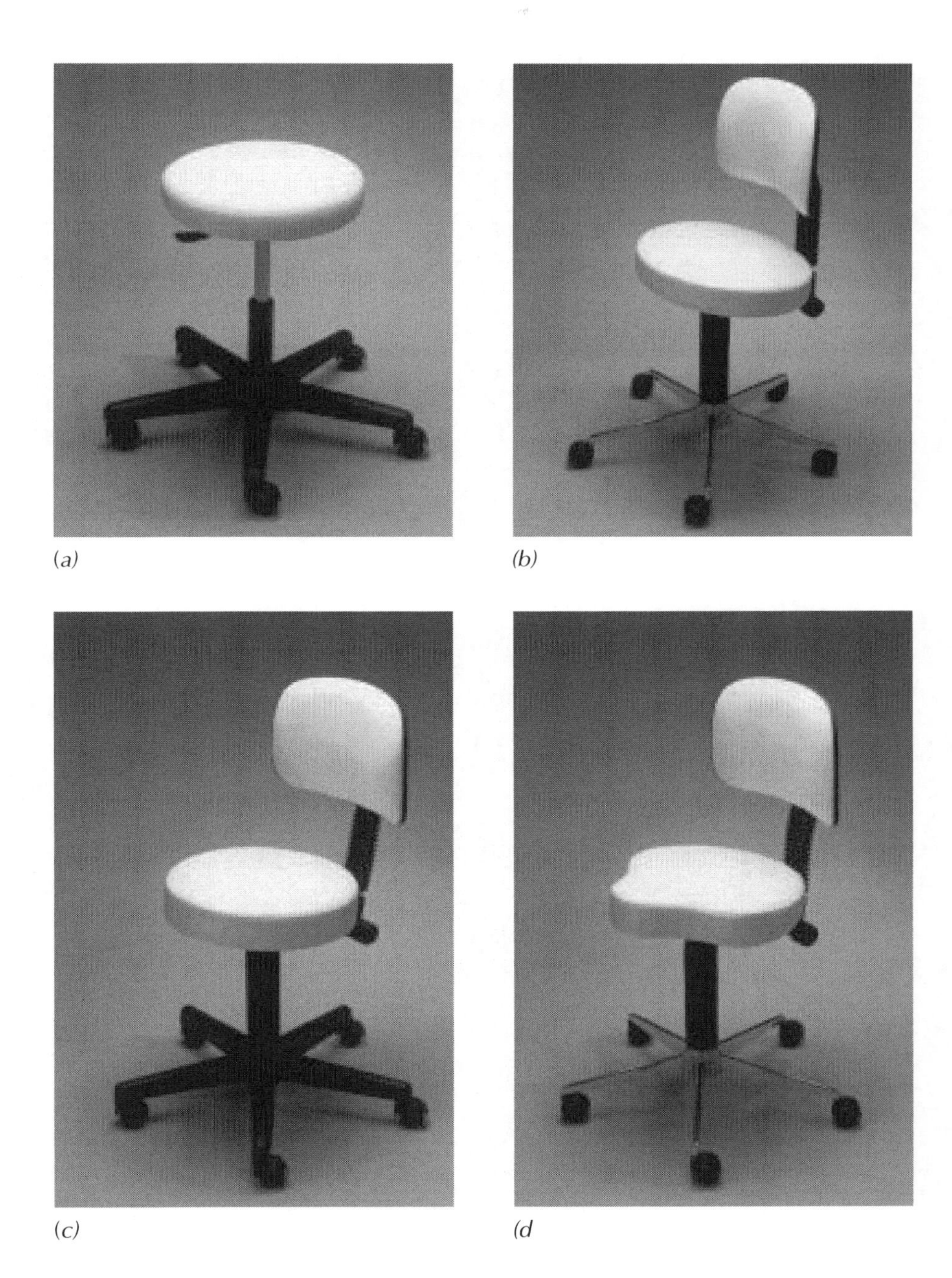

(a) *(b)* *(c)* *(d*

Figure 22.4
Operator's chair: *(a)* economy stool; *(b)* elliptical stool with back; *(c)* round stool with back; *(d)* saddle stool with back (Cosmetronic UK Ltd)

Epilation machines

There is a wide selection of electro-epilation machines to choose from. Personal preference is a good blend machine, which gives a choice of three methods of treatment – short-wave diathermy, galvanic electrolysis, and blend. These units offer flexibility according to the client's needs.

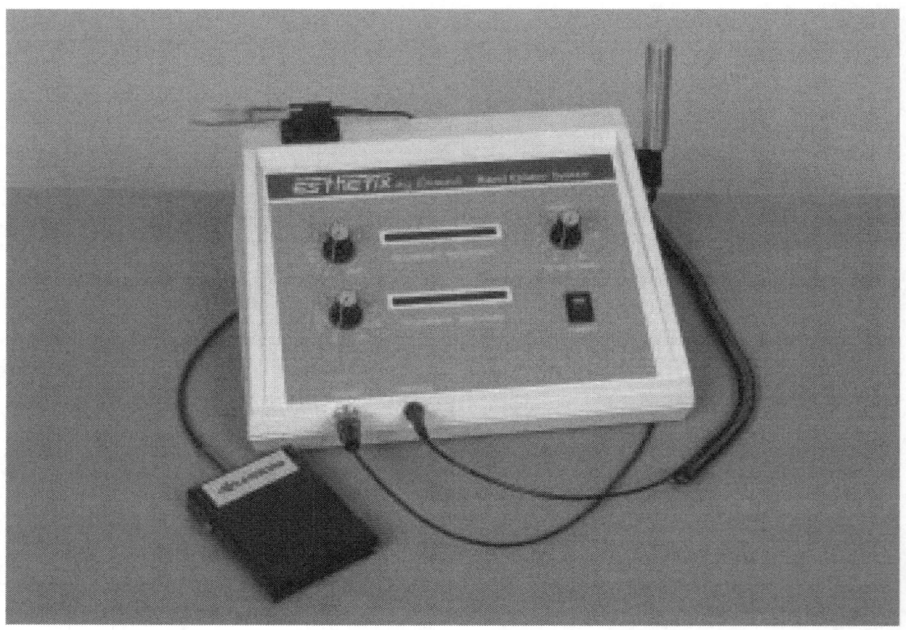

(a)

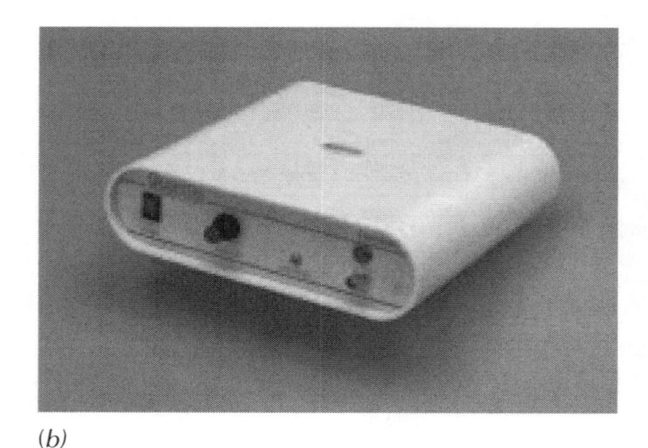

(b)

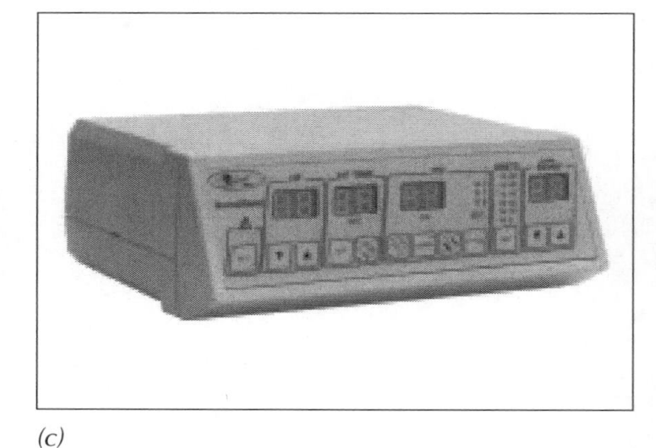

(c)

Figure 22.5 Epilation machines: *(a)* Esthetix; *(b)* Quantis (E.A. Ellison); *(c)* SonoBlend (CTI)

Machines may offer a choice of manual or computerized application. The electrolysist may choose to switch on the current flow by the use of a foot-pedal or a finger-switched needle holder. The electrolysist controls the intensity and application time of the current to the follicle.

There are now an increasing number of computerized machines coming onto the market. When using a computerized machine it is necessary for

Figure 22.6 *(top)* Carlton Professional Autoblend; *(middle)* Rita Roberts Digital Blend Unit; *(bottom)* CTI SonoBlend machines

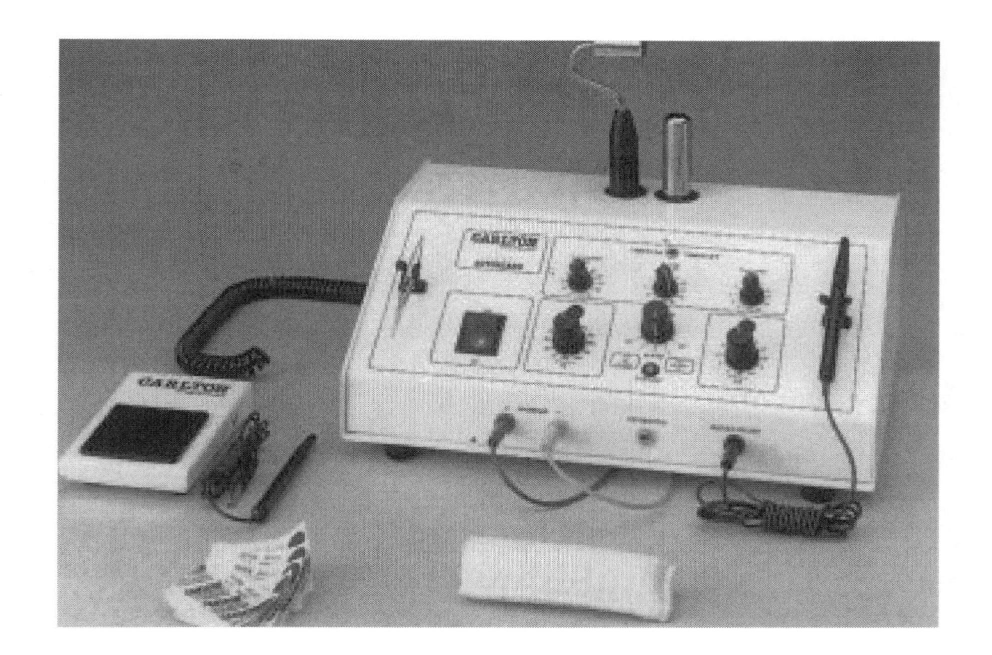

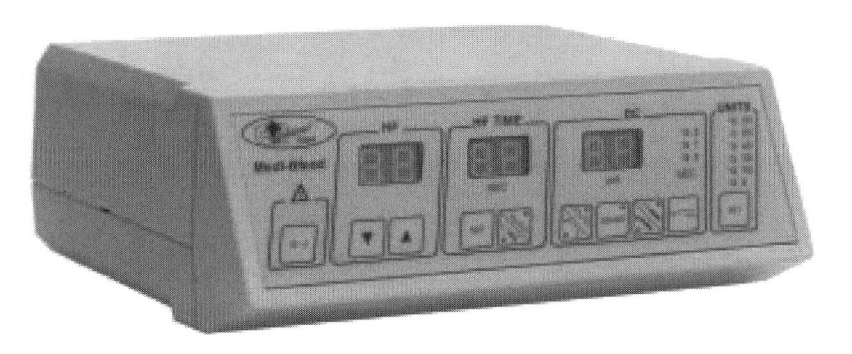

the electrolysist to be familiar with the programs available and their application. Some have the facility for the electrolysist to change methods of treatment at the touch of a button, depending on the client's needs. Treatment details and records can be stored on a floppy disk, so enabling the electrolysist to keep an accurate record of type of treatment, intensity of currents, method (e.g. blend or high-frequency), duration of treatment, area treated, cost of treatment, and number and frequency of treatments. In all instances, the machine is only as good as the electrolysist who is using it.

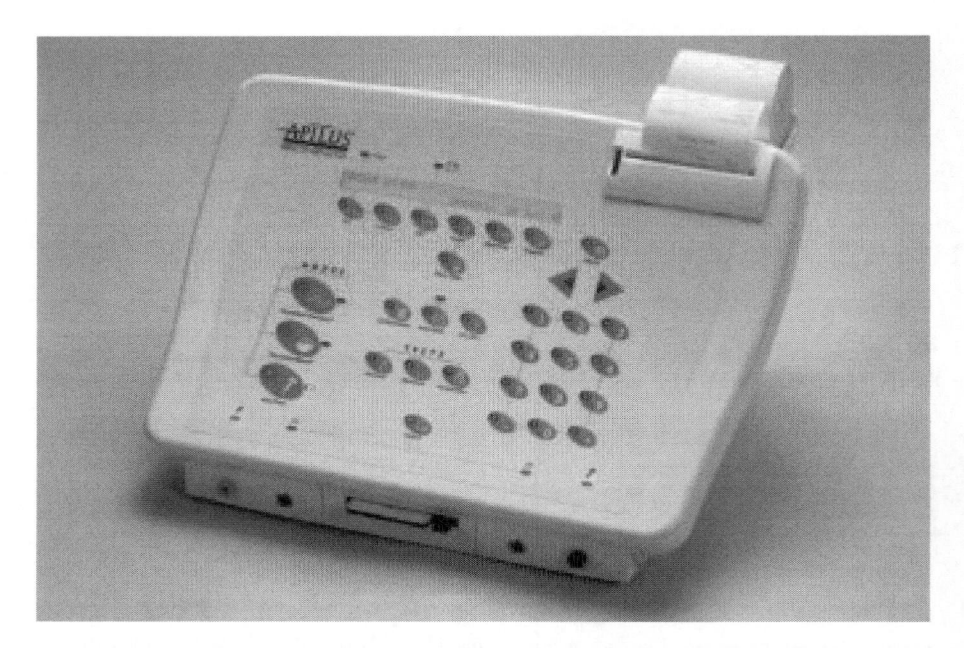

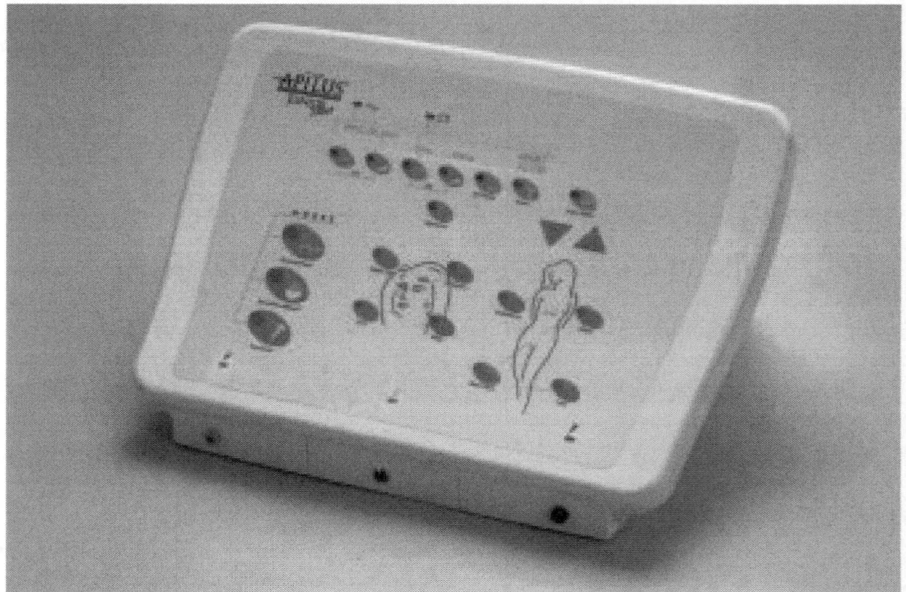

Figure 22.7 Apilus 500sx *(top)* and Junior *(bottom)* machines (House of Famuir)

The Apilus offers five different methods in one unit, these being:

- MicroFlash enables the electrolysist to perform rapid epilation of large areas. High-frequency current is applied to the follicle at a high intensity with precision to one-thousandth of a second. Owing to the speed of application the follicle can be treated with very little discomfort to the client.
- Multiplex enables the electrolysist to combine slow thermolysis with MicroFlash. This method creates porosity in the tissues, resulting in a more intense action of the MicroFlash.
- OmniBlend and Blend are ideal for treating deep, terminal hairs or distorted follicles. During this application the combination of current can be applied in one of two ways:
  (a) A combination of galvanic and high-frequency currents are applied simultaneously throughout the entire application of currents to the follicle.
  (b) Galvanic current is applied to the follicle continuously, while at the same time high-frequency is applied in pulses. This is also known as pulsing blend.
- MultiBlend combines the two techniques of blend and multiplex. MultiBlend can be set to work in automatic mode, allowing the electrolysist to establish a rhythm and so work quickly, yet efficiently.

Review questions

1. What are the points that must be considered when choosing equipment for a mobile practice?
2. Why is the choice of supplier of importance when choosing:
   (a) Equipment
   (b) Consumable materials?
3. State six points that should be considered when choosing electrical equipment for the clinic.
4. What are the requirements of the Electricity at Work Regulations 1989?
5. State the qualities of a good treatment couch.
6. What is the purpose of the magnifying lamp?
7. List the qualities of a good magnifying lamp.
8. Why does the operator's chair/stool make a difference to:
   (a) the efficiency of treatment
   (b) the comfort of the operator?
9. What are the benefits of a computerized epilation machine?

23 Advertising and public relations

When setting up in business, advertising and public relations are two valuable aids that can play a major role in attracting clients and a budget should be included in the cash-flow forecast. It doesn't matter how good you are at electro-epilation if future clients do not know of your existence. You need to get the message across – and there are a number of ways in which this can be achieved.

Advertising

When used correctly, advertising can encourage prospective clients to make contact for further information, or to book a consultation. The purpose is to let people know *what you can do for them*. It also lets people know about you, your qualifications, the services offered and the location of the clinic, or practice. Always include a contact number. The aim is to inform people of your existence but also to encourage them to come through the door. This is often the hardest step of all for many prospective clients.

For an advertisement to be effective it must be:

- Well set out and displayed.
- Carefully and honestly worded.
- Informative, but to the point.
- Tasteful and eye-catching.
- Tell the prospective client what the treatment can do for her/him.

In other words, it must get your message across clearly, without cramming too much information into a limited space. A small advertisement, placed on a regular basis, in a prominent position, is more effective than a large, expensive display that only appears once.

An *advertisement feature,* also known as advertorial, taking a quarter or half-page in the local newspaper, is often a good introduction and, when well written and presented, catches the public's eye. Half the feature could be in the form of an editorial. This is a way of bringing you, your establishment and the services offered to the attention of many prospective clients. Indeed, it is not unusual for future clients to keep such a feature for several months before finally booking an appointment.

A feature of this kind should be followed by the insertion of several smaller, yet well-placed advertisements in the same newspaper. Fortnightly insertions are often more effective than weekly appearances. When people take it for granted your advertisement will always appear, there is no urgency to make contact. However, its sudden disappearance can provide the jolt that stimulates the client into action.

The position of the advertisement is of vital importance. There is more chance of the advertisement being noticed when it is placed at the top right-hand corner of the right-hand page. The prospective client will see it as

The Sheila Godfrey Clinic

Dorridge

Telephone
01564
775546

Do any of the Following cause you embarrassment?

- **UNWANTED HAIR**
- **THREAD VEINS ON THE LEGS**
- **FACIAL RED VEINS**
- **SUN DAMAGED SKIN**
- **ACNE**
- **DEEP FACIAL LINES**
- **WRINKLES**
- **AGE SPOTS**
- **MOLES/CYSTS**
- **SKIN IMPERFECTIONS**

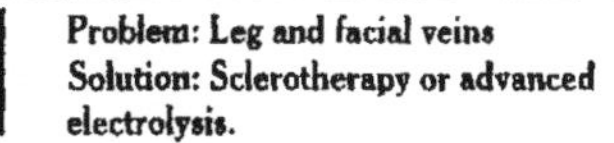

WE HAVE THE SOLUTION

From excess hair and spider veins to wrinkles and acne, The Sheila Godfrey Clinic is Solihull's first paramedical clinic and its problem-solving treatment range is as extensive as it is specialised.

Problem: Unwanted hair
Solution: Epilight or Electrolysis
The Epilight system is one step beyond lasers. The result is super-fast, long-term removal of most hair types and colour anywhere on the body for men and women. Electrolysis, established since 1875 has the longest known track record for permanent hair removal.

Problem: Acne, Pigmentation, fine lines
Solution: Preparations by M.D. Formulations
The skin-smoothing solutions can not only reverse the signs of ageing & sun damage, but also dramatically improve acne and acne scarring. Treatment is tailored to suit the needs of the individual.

Problem: Leg and facial veins
Solution: Sclerotherapy or advanced electrolysis.
The all too common thread veins are a major cause of distress for many women, yet they can be eliminated quickly and easily with Sclerotherapy for the legs with the Clinic's visiting Doctor and Shelia Godfrey's skilful needle for spider and facial veins.

Problem: Deeper lines and wrinkles
Solution: Restylane injections
Restylane is the latest alternative to collagen. Frowns, furrows and deep facial lines are smoothed away in one or two treatments. Restylane is hypo-allergenic, so a patch test is unnecessary. The effects last longer than collagen.

To find out more about how these and other treatments work book a place at Sheila Godfrey's Open Evening. Places will be limited to ensure that all who attend will be able to hear and see the presentation clearly.

To book your place and entry into the prize draw contact **Hazel Taylor on 01564 775546** (Limited spaces available)

EVENING PRESENTATION
Tuesday 22nd February 7.30pm
(at Raquetts Health Track)

Hair today - Gone Tomorrow
Hi - Tech Hair Removal

Skin Problems, Sclerotherapy

Anti - Ageing.

Prize draw

Organised by The Sheila Godfrey Clinic

Figure 23.1 Example of an advertising feature or 'advertorial'

Mrs Godfrey
DRE, FBAE, FIE, FBABTAC
International specialist in
ELECTROLYSIS
Treatment for the removal of unwanted hair and broken capillaries using disposable sterile needles
For details and confidential consultation without obligation contact:
The Electrolysis Clinic
79 Westbrook Road, Four Oaks, West Midlands B29 1QZ
Tel: 01564-060791

Figure 23.2 Example of a small display advertisement

soon as they open the page. Front and back pages are also desirable positions; however, they are usually more expensive. The Women's page, together with the Health and Lifestyle sections, also the television page, are good positions. Be careful about advertising in the personal column, as this can give the wrong impression – it depends on the quality of the newspaper concerned.

The Yellow Pages is often the first choice when someone is looking for a specific service. Yellow Pages have sections for both 'Electrolysis' and 'Health and Beauty'.

Advertisements placed in the wrong position or during a traditionally quiet time – such as Bank Holidays or main holiday periods – are a waste of money.

There are occasions when a good price can be obtained when the newspaper has not sold all the space available or they have had a late cancellation. Sometimes these late booking can be good value for money, yet at other times the return from new business does not justify the expenditure. It is therefore advisable to consider the advantages carefully before accepting a late booking.

The use of glossy magazines or television is not cost effective for the electrolysist who is in private practice; who is mobile; or who is running a small to medium sized clinic. The expense is high for a very small return.

Local businesses such as hairdressers, fashion outlets and hotels are often happy to leave your business card on display. Contact should be made through the proprietor or manager.

Plan your advertising campaign with care and allocate a budget for this purpose. In the early days, when clients are few, carefully planned and placed advertising can pay for itself relatively quickly. As the number of clients increases, the need for advertising decreases.

The role of computers

Computers are an invaluable tool for keeping track of enquiries, clients' appointments, frequency of appointments, birthdays and addresses. They also have a role in advertising. A simple newsletter can be written and professionally presented to keep clients informed about events in the clinic.

A mail shot can be organized easily from data stored on the computer, as newsletters and address labels can be produced quickly.

Web site

Time spent on the design and setting up of your own web site can lead to a steady flow of enquiries and subsequently new clients. A web site enables you to promote your services, your clinic and your own skills. The Internet and use of web sites can be an asset to any business. There are an increasing number of people who turn to the Internet for information on specific treatments.

The web site should be well designed and interesting to look at. It should create a desire for further information. Ideally, individuals who are interested in knowing more about the treatments you offer should be able to leave their name and address and details of the information they require. The web site should be updated regularly. It is an ideal medium for promoting special offers and a very powerful tool. Having set up your web site, remember to make people aware of its existence.

Public relations

Public relations are all about promoting the business through image, presentation and making contact – both socially and professionally (networking) – with key people in the locality.

Successful public relations can be achieved in many ways:

- Free editorial in the local newspaper.
- Letter of introduction to local GPs and key medical personnel.
- Talks and demonstrations to local organizations.
- Hospital work.
- Open days and evenings.
- Guest appearances on local radio and television programmes.
- Radio phone-ins.
- Leaflets in dentists' and doctors' waiting rooms.
- Networking.

Taking each of the above points in turn, how can they be achieved?

Unwanted hair causes distress and embarrassment to many people. The features editor of the women's page at the local paper may be interested in writing an article on the subject. To achieve this you must find an angle to the subject that will be of interest to readers – one that shows them the solution to a specific problem they may have. It helps if you already advertise with the paper. Ring the editor and introduce yourself and follow this with a meeting to discuss the subject. Before the meeting make sure you have all the relevant information to hand, with the important facts in writing – this helps to ensure the important facts are taken down correctly. A press release could be sent prior to the meeting to give the editor some background information. If the editor is not familiar with electrical epilation, write the article yourself. The inclusion of some before and after photographs, illustrating case histories, are always received well by editors. **However, be sure to obtain permission from the clients concerned.** Be prepared to promote yourself and your business without being too pushy. Maintain the professional image.

Another option is to send a letter of introduction to local doctors, particularly those in private practice, gynaecologists, endocrinologists and dermatologists. The letter should include details of your experience, qualifications, membership of relevant professional organizations and the location of your practice. Remember to keep the letter of introduction short and to the point.

When necessary, liaise with the client's doctor, but do *not* waste his/her time with trivial matters. You should state the name and address of the client and the nature of the problem, together with brief details of the proposed treatment. A sample letter is shown in Chapter 15, page 120.

Talks and demonstrations to local organizations such as luncheon clubs, Women's Institutes, Townswomen's Guild, Business Women's associations are a good way of introducing the subject of electrical epilation to many women who want to know more about the subject but who are too shy or embarrassed to ask. Often for these women it is a case of not knowing who to go to for treatment, or how to find out! The audience at these meetings will be looking at you for at least 30–45 minutes. This is an ideal opportunity to promote yourself and your knowledge while creating the right impression.

An open day or evening, perhaps on behalf of a charity, brings clients and members of the public into the clinic and creates interest.

A guest appearance on a local radio phone-in or television programme is a good way of introducing the subject of electrical epilation to the public. Make contact by ringing the producers or researchers of suitable programmes. These people are always looking for interesting stories and subject matter. Although programmes will not mention specific clinics it will be possible to give the name and addresses of the professional associations concerned with electrical epilation. Interested parties will then contact the association, who in turn will pass on your name – provided, of course, you are a member!

Networking refers to developing personal contacts with people who are in the position to refer business to you. There are a number of ways in which this can be achieved when networking is used correctly.

- Membership of professional associations. Colleagues from different parts of the country will often give referrals to their clients who are moving to a different area.
- Joining a networking group such as Business Network International (BNI). The aim of this organization is to provide a structured and supportive system of giving and receiving business. It does so by providing an environment in which members gain the opportunity to develop personal relationships with other qualified and skilled business people.
- Becoming a member of a 'professional women's' networking breakfast or luncheon club, where in a social environment personal contacts can be developed.
- Business cards should be carried with you at all times. You should never miss an opportunity to pass your card on to people who may wish to contact you in the future, or who may be in the position to pass this card on to a third party.

Conclusion

Promotion, public relations and advertising all contribute to the development and growth of a successful clinic or private practice. All three areas need to be constantly reviewed and updated. Well-placed advertising tells the prospective client what your treatments can do for them. Promotion is about yourself, your business, your qualifications and knowledge, and helps to create your image as well as build confidence in clients. Public relations is concerned with getting the message across through editorial, open evenings/days, talks and demonstrations etc.

There are still other things you can do to give your business every chance of being successful. For example, *the clinic and treatment rooms* should be warm, with a welcoming atmosphere and at the same time should portray a professional image. The décor should be subtle, encouraging relaxation. The surroundings should be clean and hygienic without appearing too clinical. The ambience should be one that encourages clients to enter the premises, book an appointment and return for further treatment.

Diplomas and certificates of qualifications (specific to electro-epilation), together with current certificates showing membership of professional associations, should be framed and displayed in a prominent position.

Written testimonials from satisfied clients can be placed together in a folder and placed in reception or a prominent position so that these can be shown to new clients.

The *personal appearance* of the electrolysist and the staff will make an immediate impact on the client. Untidy hair, heavy or bizarre make-up, pierced noses, lips or eyebrows, long, enamelled nails, scruffy uniform, laddered tights and dirty shoes do not give a favourable impression. A well-groomed appearance and pleasant manner will do much to create the right impression and instil confidence. Teaming this with expertise and knowledge of the subject is a sure way to gain new clients.

Know your subject well and keep up to date with new treatments on the market, how they work and whether they live up to the claims made with regard to treatment results. Clients are always looking for alternative methods to electro-epilation that do not use needles, or which may achieve a faster result. You need to be able to supply the answers.

And finally, *do what you say you will do*. Be honest with clients and do not give them unrealistic expectations with regard to treatment results and the length of time required to achieve these results.

Review questions

1 Define the following terms:
   (a) Advertising
   (b) Public relations
   (c) Promotion.
2 Explain the purpose of an advertisement.
3 Why is it important to place an advertisement in one of the following positions:
   (a) Top right-hand of right page
   (b) Television page of the local paper
   (c) Front or back page.
4 List the information that should be included in an advertisement.
5 Name three locations where business cards or leaflets can be left.

6 Define the term 'advertorial'.
7 List eight ways of promoting your business and image.
8 Define the term 'public relations'.
9 How can an open day or evening encourage new business?
10 List the advantages of giving talks and demonstrations to local organizations.

24 Professional ethics

The term 'professional ethics' refers to the code of conduct or standards of behaviour practised by the professional electrolysist in relation to:

- Clients
- Colleagues
- The medical profession
- The professional association of which the electrolysist is a member

Courtesy, honesty and integrity are all essential qualities that separate the caring professionals from the 'cowboys'.

Ethics towards clients

At all times clients should receive caring, professional treatment. Honest information should be given in relation to duration and progress of treatment. False promises or prolonged treatment for financial gain are not beneficial to the electrolysist's long-term reputation.

Clients should be able to expect and receive *confidentiality* at all times. Information relating to one client should not be discussed with another. Conversations relating to controversial subjects such as politics, religion and racial matters are best avoided.

Clients' appointments, once made, should not be cancelled or altered by the electrolysist without good reason. There are times when an event such as illness, a death in the family, failure of the electricity supply or a major crisis may prevent the electrolysist from keeping an appointment. In such instances the clients should be notified in advance, so avoiding wasted journeys and possible ill feeling. The majority of clients are very understanding in such circumstances.

The appointment book should be organized to allow for accurate time keeping so clients are not kept waiting. Running late occasionally is acceptable; to do so on a regular basis shows inefficient organization together with lack of courtesy towards the client.

Many clients are inclined to talk about their personal problems during treatment. The electrolysist is often the only person they can talk to about such matters. It is essential conversations remain confidential and are *never* discussed with a third party. In these circumstances the electrolysist should not offer personal advice, since this could give rise to a number of difficulties at a later date.

The client's best interests should always be of prime consideration.

Ethics concerning colleagues

A true professional does not attempt to poach clients from colleagues, and does not speak disparagingly about another electrolysist's standard of work.

Occasions arise when an electrolysist will attend to a colleague's clients on a temporary basis, for example, during holiday periods or times of illness. When this situation occurs, the original treatment plan should not be altered in any way without consultation with the colleague concerned.

A minority of clients exist who tend to flit from one clinic to another, or who make appointments with two electrolysists. Once this situation has been brought to the attention of either operator, it must not be allowed to continue, in order to safeguard the electrolysists concerned and to prevent the over-treatment of the client's skin. Such situations prevent accurate record keeping and monitoring of treatment progress.

Ethics towards the medical profession

Clients expect electrolysists to be able to recognize and pass an opinion on any number of skin conditions and medically related matters. It is well known that many skin lesions demonstrate similar characteristics, some of which may be benign, whereas other may be malignant, but it is not for the electrolysist to diagnose. Electrolysists should only pass an opinion on subjects in which they are qualified. There will be times when treatment should not be given without first obtaining written agreement from a medical practitioner.

Electro-epilation treatments can be divided into three categories: those that cover cosmetic purposes; those that are hormone-related but do not require medical liaison, e.g. menopause; and those for which medical referral is necessary.

The initial consultation may indicate the possible existence of a condition that requires medical investigation prior to electro-epilation, or an established disorder such as diabetes or epilepsy. In such circumstances the client's GP must be consulted before any treatment is given.

It is essential the electrolysist is aware of when referral to the client's doctor is advisable. Equally, busy doctors soon become irritated with unnecessary communication on irrelevant matters. When a medical practitioner agrees to electro-epilation treatment, progress reports may be sent to that practitioner as and when necessary. This not only keeps the doctor informed but also helps to build a good working relationship between the medical and electrolysis professions.

Professional associations

There are several associations in the UK for electrolysists and beauty therapists. These include the British Association of Beauty Therapy and Cosmetology; the Guild of Professional Beauty Therapists; and the Society of Professional and Holistic Therapists.

There are only two specialist associations for qualified electrolysists. These are:

- The British Association of Electrolysists

BRITISH ASSOCIATION OF ELECTROLYSISTS LTD
Electrolysis using needle epilation is the only acceptable procedure for Permanent Hair Removal by BAE Members

- the Institute of Electrolysis

THE

TREATMENT OF

SUPERFLUOUS HAIR

BY

ELECTRICAL EPILATION
(DIATHERMY) GALVANISM
OR THE BLEND

This is an explanatory leaflet prepared by the INSTITUTE OF ELECTROLYSIS LIMITED, in the interest of its members' patients and the general public requiring professional treatment for superfluous hair.

It must not be copied wholly or in part without the written permission of the Council of the Institute of Electrolysis Limited. ©

Figure 24.1 Front cover of explanatory leaflet prepared by the Institute of Electrolysis

Both organizations have similar aims and objectives, namely, raising and maintaining standards together with promotion of the professional status of electrolysis.

Membership of an association provides many benefits for the practising electrolysist, including:

1 Conferences and meetings for interchange of ideas and knowledge between colleagues
2 Lectures and demonstrations
3 Point of contact for members of the public seeking a list of qualified practitioners
4 Promotion of electro-epilation through publicity in national magazines, newspapers and reference libraries
5 Refresher and advanced epilation courses
6 Preferential insurance rates

The Institute of Electrolysis makes available an explanatory leaflet (see Figure 24.1) which aims to give a brief explanation of the types of treatment available and alleviate any concerns actual or prospective clients may have.

The associations, quite rightly, expect members to maintain high professional standards and to adhere to their codes of conduct. Each association holds an annual general meeting (AGM) at which all business matters and information relating to the running of the association are presented to the membership.

Both associations expect high standards with regard to treatment and code of conduct as outlined earlier.

It is the author's opinion that any electrolysist who is serious about his/her career will not only gain many benefits from belonging to a professional association, but by taking an active part will also be able to make a valuable contribution to the profession as a whole. For those students and qualified electrolysists who are interested in becoming a member, a list of professional associations and contact addresses is given at the end of this book.

Review questions

1 Define the term 'professional ethics'.
2 State why clients' appointments, once made, should not be changed without good reason.
3 Explain why clients should not be permitted to book appointments in more than one salon.
4 When should an electrolysist refer a client to the doctor?
5 Name the two British specialist associations.
6 List the benefits of belonging to a professional association.

25 Case histories

The following case histories relate to clients who have attended the author's clinic with successful results.

Case history 1

A 58-year-old woman with a heavy growth of terminal hair on the chin, neck, jaw line and upper lip, attended the clinic for an initial consultation. Previous treatment with an unqualified operator had not been successful. This client was disenchanted with electro-epilation. However, desperation concerning her problem led her to try again, this time having obtained a list of qualified practitioners through the Institute of Electrolysis.

The initial consultation revealed that surgery for the removal of fibroids and abscesses from the womb had taken place. There was no history of menstrual irregularities or hormone disturbance other than those associated with the menopause, which had not caused any problems. The cause of the hair growth appeared to be a genetic predisposition relating to sensitivity of follicles to circulating androgens. Mechanical interference, including regular tweezing and waxing, aggravated the condition.

The client was difficult and argumentative during the consultation. She was adamant she wanted to attend the clinic daily for treatment. A compromise was reached of 20 minute sessions three times a week, working in rotation on the chin and jaw line, upper lip and neck. In this way the client was given peace of mind, yet each area was given sufficient healing time before receiving further treatment. Had this client not been allowed treatment so regularly, there is no doubt she would have attended two clinics for treatment without keeping either informed. Psychology had a large part to play in this client's treatment programme.

Blend treatment was the method chosen for the chin, neck and jaw line, with short-wave diathermy being used on the upper lip. During the course of treatment it was noted regrowth was considerably less on the areas where blend had been used. Treatment sessions were for 20 minutes, three times weekly for six months, dropping to twice weekly for five weeks, then weekly thereafter. Appointments gradually became less frequent as the condition improved. A successful result was achieved over a period of 20 months.

During this period the client's attitude and personality changed from being aggressive, rude, argumentative and difficult to please, to pleasant and outgoing, with a lively sense of humour. Self-esteem improved, together with a remarkable change in appearance. This client joined Weight Watchers, and lost 15 kilos; had her hair restyled; and took far more interest in clothes. Her husband was delighted with the change in his wife. She never told her husband she was having electro-epilation treatment, because she was embarrassed.

Case history 2

A 61-year-old woman, with one son aged 26, had a problem with excessive hair growth that started at the age of 17 after she had received progesterone injections. Not only was this client lacking in confidence, but she was also shy and reserved. The hair growth was present on the chin, sides of face, neck, jaw line and upper lip. The client spent 25 minutes waxing and tweezing on a daily basis. The skin became sore and the hair became increasingly coarse. Her son, aware of his mother's distress, contacted the British Association of Electrolysists for a list of qualified electrolysists.

Initially, treatment was given with short-wave diathermy to clear the area quickly, using a size 004 needle. Treatments sessions were for 45 minutes on a weekly basis. After four weeks, treatment was changed to weekly sessions of one hour using blend. Over a period of 13 months, treatment sessions gradually became shorter until sessions were of 15 minutes duration. The needle size changed from 004 to 003, and eventually to 002. The length and frequency of the treatments were then determined by the appearance of regrowth. Treatment was completed after a period of 25 months.

When this client first attended for treatment she would rush into the clinic and found it very difficult to make eye contact with any member of staff. Her self-confidence was non-existent, and her life revolved around her son. As her hair growth problem decreased, she became more out-going and interested in life. She was encouraged to take up golf as a hobby, which soon became an overwhelming passion.

She has since recommended several of her friends to the clinic for treatment. Her words on more than one occasion were: 'No longer having to wax and tweeze on a daily basis has given me a new lease of life. Electrolysis has changed my life – I cannot believe the difference it has made to me ...'

Several years later, this client suffered a severe heart attack and spent two weeks in intensive care. When she began her road to recovery her comment to her son was that her first thought when she was moved from the intensive care unit was: 'Thank god I had electrolysis when I did. I did not have to suffer the indignity of lying in hospital with a hairy face.'

Case history 3

This case concerns a 19-year-old university student. She first attended the clinic at the age of 16, with fine, soft dark hair on the upper lip, which caused a shadow. This was treated with short-wave diathermy using a size 003 needle, thinning out the number of hairs present. Ten sessions of 15 minutes duration, given at two-to-three-weekly intervals, achieved a successful result. This client returned three years later in distress. She had gained weight steadily, coarse hair growth had appeared in a masculine pattern on the face, and acne-type lesions were present. The skin had become excessively oily, and menstruation was irregular. This client was referred to her GP, who in turn referred her to a gynaecologist.

Polycystic ovary syndrome (Stein–Leventhal syndrome) was diagnosed. Treatment involved hormone therapy in conjunction with blend electrolysis over a period of nine months. This particular situation was diagnosed and dealt with at the early stages of development. Consequently, hormone therapy was able to deal with the underlying cause relatively quickly. Blend dealt with the existing hair growth, which had not been aggravated by mechanical interference. Blend was given at fortnightly intervals for a period of fifteen minutes during the university holidays only. Treatment was completed in 15 months.

The polycystic ovary syndrome flared up periodically over the next ten years. The client moved to London, where she attends for further electro-epilation when necessary. The client has accepted the hair growth could be an ongoing problem, owing to the underlying cause. However, this no longer causes her distress due to the fact she can deal with the intermittent hair growth with electro-epilation.

Case history 4

An Asian solicitor, aged 34 years, attended the clinic for a second opinion, having had short-wave diathermy with another electrolysist. Deep pigmentation marks were present where she had received treatment to the eyebrows, chin and upper lip. There were also noticeable pit-marks around the follicle opening. Without doubt, the pigmentation marks had occurred as a result of too much heat being applied too close to the skin's surface. The pit marks could have been the result of deep insertions, where the needle may have pierced the base of the follicle, so applying current to the underlying tissues.

The hair texture was generally fine, with the exception of some darker, coarse hairs on the chin, sides of upper lip and eyebrows. The skin was fine in texture, with good moisture content.

Photographs were taken prior to further treatment being given. The client was asked to sign a statement relating to the presence of the scars and pigmentation. The skin was allowed to recover for a period of three months before commencing further treatment.

Treatment, using blend, was given at weekly intervals, either to the eyebrows or to the chin and lip. The treatments were rotated to make sure each area was treated fortnightly. Blend was the preferred method, so preventing a build up of heat within the skin structures. Treatment sessions were of 20 minutes' duration.

The skin was prone to react badly when the client was tired, stressed or when the weather was hot and humid. The reaction would show in excessive oedema/swelling around the follicle and the gradual appearance of erythema. When this happened, treatment was stopped immediately and cool compresses applied to the area.

The needle was changed to the Ballet insulated needle. This had a beneficial effect, with less skin reaction.

Over a period of 14 months the pigmentation marks gradually faded, until they had disappeared. The pit marks decreased in size significantly, but were still present at the time of writing.

NB: Asian skin is very sensitive to heat and will mark easily if the electrolysist does not take care when giving treatment.

Case history 5

This was a male client aged 26 years, with an English mother and Jamaican father. This client was concerned with hair growth along the cheek bones. The skin became sensitive as a result of constant shaving. This was aggravated by the fact this client worked outside in all weather conditions.

The skin acquired a tan easily and was prone to developing pigmentation marks. The hair was very coarse with straight follicles.

Initially treatment was given every two weeks for 30 minutes, divided into 15 minutes for each cheek. Blend was used due to the texture of the hair, the moisture content and the tendency to develop pigmentation marks. After three months, appointments were spaced further apart, frequency depending on the appearance of regrowth.

A gold 004 needle was used at the beginning of treatment, but was changed to gold 003 as the hair became finer in texture.

Case history 6

This 29-year-old sales executive had dark, fine hair on her upper lip and sides of the face. The skin was dry, with the hairs lying very close to the skin's surface.

The hair growth was due to a hereditary disposition with no hormonal causes. Treatment by short-wave diathermy was given on a weekly basis initially, using a 002 insulated needle. Treatment sessions were 15 minutes in length. Over a period of three years, treatment sessions were gradually spaced further apart until the client returned every two months.

Psychologically, this client had difficulty in accepting the hair growth was clearing and treatment was no longer necessary. This is an instance where a photograph taken at the beginning of treatment would have been of great help in convincing the client the hair growth had decreased substantially.

Case history 7

This client was a 49-year-old magistrate who had noticed the appearance of dark hair on the upper lip and chin, corresponding with the start of irregular periods, when she was 47. Hair had been removed regularly by waxing and tweezing. Growth was of medium texture, with distorted follicles. The client did not need hormone replacement therapy and was not being prescribed any form of medication.

Treatment was successfully carried out over an 11-month period using blend. Initially, 20 minute sessions were given on a weekly basis, then fortnightly, progressing to three visits during the last three months.

Case history 8

A 10-year-old Asian girl attended the clinic with her mother. The girl was very distressed because of the dark hair growth on her upper lip. The hair was due to a combination of racial predisposition and hormonal changes. After consultation with the mother initially and the mother and daughter together, a short course of electro-epilation with short-wave diathermy was planned. Treatment sessions lasted between 5 and 15 minutes depending on how this girl felt on the day. Treatment sessions were given every 10–14 days over a period of 4 months. The hair growth was reduced to an acceptable level and the girl became less conscious of the hair growth and more relaxed.

Case history 9

A 9-year-old Asian girl attended the clinic with her mother. A dense, heavy hair growth was very noticeable on the upper lip. This girl was being bullied and had started to refuse to go to school. She was extremely distressed. The hair growth was due to racial predisposition.

The treatment procedure was similar to Case 8. Amnitop was applied to the skin before treatment. Treatments were given over a 7 month period, with 2 weeks in between each appointment. At the conclusion of treatment this young girl was no longer being teased and bullied and was able to enjoy joining her friends at school.

In both Cases 8 and 9 treatment was given after in-depth consultations with the parents. (The father of the 9-year-old was in the medical profession, which made the decision much easier.) All facts were considered carefully. In both instances it was agreed that the psychological effects of the very noticeable hair growth were detrimental to the girls' self-esteem. The first girl suffered from embarrassment and lack of confidence, the second suffered the additional pressure of bullying. In both instances the written consent of the parents was obtained.

This small sample of case histories shows that clients differ widely, even though the underlying cause of their problem may be the same. Clients vary in their response and commitment to treatment. Every course of treatment must be planned to suit the needs of the individual, with the client's long-term well-being and best interests being of prime importance.

Blend or short-wave diathermy and in many instances a combination of both, are the preferred methods of treatment. Pure galvanic electrolysis is too slow and time-consuming.

Glossary

Adrenal glands are situated one above each kidney. They produce a number of hormones which include adrenaline (US name *epinephrine*) and steroids.
Advertising the aim of advertising is to inform people of services and products available.
Acid mantle a fine acidic film of sebum and sweat found on the surface of the skin. Its function is to inhibit the growth of bacteria.
Acne due to a defect in the sebaceous glands which leads to over production of sebum. It is primarily androgen induced and may indicate hypersensitization of sebaceous glands to circulating hormones.
AIDS (acquired immune deficiency syndrome) which develops as a result of infection by the human immune deficiency virus.
Allergy hypersensitivity to a substance which causes the body to react to any contact with it in an adverse manner.
Anagen active stage of hair growth, where lower follicle is rebuilt and new hair is formed.
Anorexia nervosa psychological disorder, characterized by fear of becoming fat and refusal of food. Normally affects young adolescent girls. Involves the nervous, endocrine and digestive systems.
Antiseptic chemical which inhibits or destroys the growth of micro-organisms on living tissue.
Aseptic free from organisms capable of causing disease.
Asexual hair growth not governed by hormones, e.g. scalp, eyebrows, eyelashes.
Autoclave piece of equipment used to sterilize instruments by steam at temperatures in excess of 100° centigrade.
Benign harmless. Non-cancerous.
Blend epilation the application of direct current and high frequency to the hair follicle simultaneously.
Boil staphylococcal infection of the hair follicle.
Catagen second stage of hair follicle growth cycle. Follows anagen. Hair separates from dermal papilla; club hair is formed. Lower follicle begins to shrivel and collapse.
Club hair develops during catagen. The bulk of the hair dries out and becomes brush-like. The club hair is held in the follicle by the cells of the inner root sheath.
Cauterization occurs when a high intensity of high frequency is passed into the tissue. Moisture vaporizes and tissue becomes dry.
Chemical depilatories chemical preparations which are applied to the skin in order to dissolve the hair. Temporary method of hair removal.
Chloasma patches of increased pigmentation, usually seen on the face during pregnancy. It may also occur during the menopause.
Comedone collection of sebum, keratinized cells and certain waste substances which accumulate in the entrance of a hair follicle.

Consultation the initial meeting between client and electrolysist, which should form the basis of a professional relationship.
Contraindication the presence of any condition which indicates, or shows that electrical epilation should not be carried out.
Cross-infection the transfer of infection from one person to another.
Dermis lies under the epidermis and is the largest layer of the skin. It contains blood, lymph vessels and nerves.
Depilatory waxing commercially prepared product which is used for the temporary removal of hair root and bulb from the follicle.
Dermatology the study of the skin and its diseases.
Diabetes mellitus a condition which occurs when the pancreas fails to produce sufficient insulin.
Disinfectant a chemical agent which destroys micro-organisms but not usually bacterial spores.
Double depression giving two bursts of current to the follicle during one insertion by depressing the button or foot pedal twice.
Eczema and dermatitis are both terms which may be used to describe the same condition, which varies from a mild to an inflammatory state.
Electro-epilation destruction of hair and follicle by means of an electric current which may be high frequency, galvanic current or a combination of both currents.
Endocrine gland ductless glands situated at specific sites on the body, which secrete hormones directly into the bloodstream, e.g. pituitary gland, thyroid gland.
Epidermal cord slender cord of hair germ cells which enables the retreating follicle to maintain contact with the dermal papilla.
Epidermis protective outer layer of the skin which consists of five layers.
Epilepsy a condition of the nervous system due to disturbance of the brain's electrical activity, which results in convulsive fits.
Epileptic fit temporary disruption to the normal electrical activity of the brain.
Erythema superficial reddening of the skin due to temporary increase of localized blood supply.
Ethylene oxide used as a sterilizing agent.
Fainting temporary loss of consciousness brought about by the reduction of the blood supply to the brain.
Feeder vein refers to the larger vein which supplies smaller capillaries with blood. Term used in association with the treatment of telangiectasia.
First aid immediate assistance given to a casualty in the event of an emergency situation or accident.
Flash technique application of very high intensity of high frequency to the follicle for a fraction of a second.
Gamma irradiation gamma radiation – electro magnetic radiation used for sterilization purposes, for example epilation needles.
Glutaraldehyde chemical preparation used to destroy vegetative bacteria, spores and fungi. The life span of activated glutaraldehyde varies between 14 and 28 days.
Haemophilia an inherited familial condition in which there is excessive bleeding from an injury due to a defect in the blood clotting mechanism.

Hair keratinized structure which grows out of the hair follicle.
Hair follicle sac-like indentation of the epidermis which grows down to the subjacent dermis.
Hair germ consists of undifferentiated cells, which produce new hair when stimulated by circulating hormones and enzyme action.
Heating pattern shape of the heated area of tissue surrounding the needle during the application of high frequency.
Hepatitis inflammation of the liver. An acute infectious, viral disease.
Herald patch refers to the first lesion to appear on the skin at the onset of *pityriasis rosea*.
Herpes simplex acute viral disease characterized by formation of clusters of watery blisters.
Herpes zoster technical name for shingles. It is caused by a virus which is related to the chicken pox virus.
High frequency is produced from an oscillating alternating current of very high frequency and low voltage ranging from 3–30 MHz or 3–30 million cycles per second.
High-frequency field the heating pattern radiating from an epilation needle, connected by a wire to a high-frequency oscillator. (Shortwave diathermy machine.)
Hirsutism masculine pattern of hair growth in women, which is normal in men, caused by increased sensitivity of hair follicles to circulating hormones or by an endocrine disorder.
HIV Human immune deficiency virus which interferes with the immune system, reducing the body's ability to cope effectively with disease or infection.
Hormone complex chemical substance produced by the endocrine glands, which stimulates or inhibits the action of specific glands, organs or tissues.
Hormone replacement therapy (HRT) is the use of natural hormones to replenish the decreased hormone levels which occur during the menopause or after a total hysterectomy.
Hypertrichosis generalized overgrowth of vellus and terminal hairs. Occurs in both sexes. This condition is not hormone dependent but may be due to racial or genetic predisposition.
Hypothalamus part of the mid brain. Situated between the thalamus and the pituitary gland.
Impetigo superficial, contagious, inflammatory disease caused by streptococcal and staphylococcal bacteria.
Infundibulum funnel shaped opening to follicle.
Keloid excessive formation of scar tissue at the site of an injury to the skin.
Keratin hard horny substance made up of carbon, hydrogen, sulphur, oxygen and nitrogen; occurs in hair, nails and stratum corneum.
Keratogenous zone area where keratinization takes place in the hair follicle.
Lye is the popular name given to sodium hydroxide.
Malignant severe or threatening to life. The term is applied to any virulent condition which tends to go from bad to worse. Often refers to a cancerous condition.

Menopause the period at which a woman's menstrual cycle stops.
Menstrual cycle usually a 28 day hormonal cycle commencing at puberty, normally ending at the menopause, with a natural interruption during pregnancy.
Moisture gradient refers to the moisture content in the different layers of the skin.
Oedema accumulation of water/fluid in the tissues.
Ovary small endocrine gland which forms part of the female reproductive organs. The ovary produces ova (eggs), oestrogen and progesterone.
Pain threshold the level of stimulation to free nerve endings that can be comfortably tolerated by the individual.
Pancreas large elongated glandular organ with exocrine and endocrine functions. Situated in the curve of the duodenum, behind the stomach.
Parathyroid glands are four small glands embedded in the surfaces of the thyroid gland. Their hormones are responsible for maintaining calcium levels in the blood.
Pilary canal consists of the upper third portion of the outer root sheath, which extends above the entrance of the sebaceous gland.
Pilo-sebaceous unit formed from the hair follicle and the sebaceous gland.
Pituitary gland known as the master gland of the endocrine system due to its influence on the other endocrine glands. Produces trophic hormones.
Pityriasis rosea self limiting skin disease with no known cause. Normally runs its course within six weeks.
Plucking mechanical removal of an individual hair by means of a pair of tweezers. Also known as tweezing.
Premenstrual tension (PMT) is the name given to a collection of mental and physical symptoms, due to hormonal changes, which occur for up to ten days before menstruation.
Professional ethics code of conduct or standards of behaviour practised by the professional electrolysist in relation to clients, colleagues, the medical profession and professional associations.
Psoriasis chronic, inflammatory, non-contagious condition of the skin.
Puberty the period of change from childhood to adolescence when the sex glands become active.
Public relations (PR) is concerned with promoting a business through image, presentation and personal contact.
Recovery position is used to keep a casualty's air passages open and free from obstruction, particularly when the casualty is unconscious.
Rhinophyma a condition of the nose which shows thickening of the skin and enlargement of skin tissues. Usually a high colour is present.
Rosacea a chronic skin disease of the face; characterized by redness and the formation of pustules. Coarsening of the skin also occurs.
Sebaceous cyst round, nodular lesion with a smooth shiny surface which develops from a sebaceous gland.
Sebaceous glands are attached to hair follicles. Their function is to produce sebum.
Sepsis presence of infection due to germs or micro-organisms.
Shock disturbance of the emotions, or state of bodily collapse due to severe bleeding, burns, fright, unpleasant news or electrical shock.

Sodium hydroxide caustic chemical substance, produced during galvanic electrolysis. Destroys tissue by chemical action. Also known as lye.
Spider naevus collection of telangiectasia which radiate from a central papule.
Sterilization process used to achieve total destruction of all living organisms and spores.
Sudoriferous gland more commonly referred to as sweat gland. Found all over the body. The function of these glands is to produce sweat which regulates body heat and eliminates waste products.
Sugaring commercial substance based on lemon juice, sugar and water which removes hair in a similar manner to waxing when applied to the skin.
Telangiectasis dilation of small blood vessels in the skin. Often referred to as broken veins, red veins, thread veins, or dilated capillaries.
Telogen final stage of hair growth cycle, which follows on from catagen.
Thyroid gland largest endocrine gland situated in the neck. Its secretion controls metabolism and body growth.
Transsexual a person who feels strongly that he/she is trapped in the wrong biological body.
Units of lye tenths of a milliamp × times in seconds = units of lye. i.e. one tenth of a milliamp of dc flowing for one second produces one unit of lye.
Urticaria development of red wheals which may later become white. Lesions appear rapidly and disappear within minutes over a number of hours.
VAT refers to value added tax, which is levied on most business transactions which take place within the European Union (EU), UK and Isle of Man.
Vitiligo name used to describe lack of pigmentation in the skin.
Warts viral infection. Well defined, benign tumours which vary in size.

Bibliography

Anatomy and Physiology, Evelyn Pearce (Faber & Faber)
Gray's Anatomy (Classic Collector's edition, Churchill Livingstone)
Ross and Wilson's Anatomy and Physiology, Kathleen Ross and J.W. Wilson (7th edition, Churchill Livingstone)
Advances in Biology of Skin and Hair Growth, Montagna and Dobson (Pergamon Press)
Essential Notions of Black Skin, Humbert Pierantoni
Structure and Function of the Skin, Montagna and Ellis
De Launey and Land's Principles and Practice of Dermatology, 3rd edn, Harvey Rotstein (Butterworth Heinemann, 1993)
The Biology of Hair Growth, Montagna and Ellis
Human Hairgrowth in Health and Disease, David Ferriman
Hair Research 1981, Status and Future Aspects, edited by C.E. Orfanos, W. Montagna and G. Stuttgen
The Cause and Management of Hirsutism, edited by R.B. Greenblatt, V.B. Mahesh and R.D. Gambrell
The Hirsute Female, R.B. Greenblatt
Basic Knowledge of Esthetiques, Humbert Pierantoni
The Principles and Practise of Hairdressing, Leo Palladino
Endocrinology, John C. Small, Michael Clarke-Williams (Heinemann)
Hormones and the Body, A. Stuart Mason (Pelican)
Managing the Menopause II – A Guide to HRT, Cathy Read (edited David Sturdee)
No Change, Wendy Cooper (1990)
Dermatology, P. Hall-Smith, R.I. Cairns and R.L.B. Beare (2nd edn, CLS)
Dermatology (updated edition), Lionel Fry
Lecture Notes on Dermatology 1990, Robin Graham-Brown and Tony Burns (Blackwell)
Roxburgh's Common Skin Diseases, John D. Kirby (15th edition, H.K. Lewis, 1986)
Acne Update (Postgraduate Series), W.J. Cunliffe (Update Books)
Eczema and Psoriasis Update (Postgraduate Series), R.H. Champion (Update Books)
Body Shock, The Truth about Changing Sex, Liz Hodgkinson
Guidelines for Transsexuals Male to Female and Female to Male, SHAFT
Speech Pathology Considerations in the Management of Transsexualism, Jennifer Oates and Georgia Dacakis
Principles of Electrology and Shortwave Epilation, edited by Arthur Mahler (The Instantron Company, 1986)
Science for the Beauty Therapist, John Rounce (Stanley Thorne)
The Blend, Michael Bono (1995)
Clayton's Electrotherapy, Sheila Kitchen and Sarak Bazin (10th edition, W.B. Saunders)
Electrolysis Exam Review, John Fantz

Electrolysis, Thermolysis and the Blend, Arthur J. Hinkle and Richard Lind
The Fantz Guide to Electrolysis, John Fantz (Laurel Publications, 1983)
Health and Beauty Therapy – A Practical Approach for NVQ Level 3, Dawn Mernagh and Jennifer Cartwright (Stanley Thorne, 1995)
CIBTAC/NVQ/SVQ Guide Level 3– Electrical Epilation (1994)
First Aid Manual, St John Ambulance, The British Red Cross Society, St Andrew's Ambulance Association (7th edition, 1997)
Hygienic Skin Piercing, Dr Noah (1988)
Sterilisation and Hygiene, W.G. Peberdy (Stanley Thorne, 1988)
The Beauty Salon and Its Equipment, John Simmons (Macmillan, 1989)
Croner's Health and Safety at Work (1990)
Good Practice in Salon Management, Dawn Mernagh-Ward and Jennifer Cartwright (Stanley Thornes, 1997)
Maintaining Portable Electrical Equipment in Offices (HS (G) Books, 1994)
Memorandum of Guidance on the Electricity at Work Regulations 1989 (Health and Safety Executive, HMSO)
COSHH: A Brief Guide to the Regulations (Health and Safety Executive, HMSO, 1999)
Data Protection Act 1984, The Guidelines 3rd Series, November 1994, Office of the Data Protection Register
Department of Health and Social Security, Leaflets FB30, N140, N141
Employing People, ACAS Handbook for Small Firms (1996)
Employment Legislation – Individual Rights of Employees. A guide for employers and employees. PL 716 (REVG 7). Department of Trade and Industry (April 2000)
Guide for Small Businesses, National Westminster Bank plc
Inland Revenue Publications, Leaflets IR57, IR109
Income Tax: Guide to Self-Assessment for the Self-Employed, Inland Revenue Publications
The VAT Guide, HM Customs and Excise

Professional associations

UK associations

British Association of Electrolysists Ltd
40 Parkfield Road
Ickenham
Middlesex
UB10 8LW
Telephone: 0870 1280477 Fax: 0870 133047
e-mail: sec@baeltd.fsbusiness.co.uk

The Institute of Electrolysis
PO Box 5187
Milton Keynes
MK4 2ZF
Telephone 01908 521511
www.electrolysis.co.uk
e-mail: institute@electrolysis.co.uk

Membership of the above specialist associations may be gained by demonstrating practical skills to a board of examiners and successfully sitting a theory paper.

British Association of Beauty Therapy and Cosmetology
BABTAC House
70 Eastgate Street
Gloucester
GL1 1QN
Telephone 01452 421114
www.babtac.com
e-mail office@babtac.com

The Guild of Professional Beauty Therapists Ltd
Guild House
P.O. Box 310
Derby
DE23 9BR
Telephone 01332 771714 Fax 01332 771742
www.beauty-guild.co.uk

Overseas associations

American Electrology Association
106 Oak Ridge Road
Trumball
CT0 6611

Canadian Organization of Professional Electrologists
170 St George Street
Suite 408
Toronto
Ontario M5R 2M8
Canada

Federation of Canadian Electrolysis Associations
PO Box 1513
Station B, Mississauga
Ontario L4Y 4G2
Canada

International Guild of Professional Electrologists Inc
Professional Building, Suite C
202 Boulevard
High Point
North Carolina
USA 27260

Society of Clinical and Medical Electrologists Inc
PO Box 52
Killeen
Texas
USA 76540

Examination boards in the UK

British Association of Electrolysists Ltd
40 Parkfield Road
Ickenham
Middlesex
UB10 8LW
Telephone 0870 1280477 Fax: 0870 133047
e-mail: sec@baeltd.fsbusiness.co.uk

The Institute of Electrolysis
PO Box 5187
Milton Keynes
MK4 2ZF
Telephone 01908 521511
e-mail: institute@electrolysis.co.uk
website:

The above boards are concerned specifically with electrical epilation. The following boards examine subjects relating to beauty therapy in conjunction with electrical epilation.

Confederation of International Beauty Therapy & Cosmetology (CIBTAC)
BABTAC House
70 Eastgate Street
Gloucester
GL1 1QN
Tel: 01452 42114
e-mail: office@babtac.com

City and Guilds of London Institute
76 Portland Place
London W1N 4AA
Telephone 0207 1278 2468

EDEXCEL
Stuart House
32 Russell Square
London
WC1B 5DN

Vocational Training Charitable Trust
46 Aldwich Road
Bognor Regis
West Sussex
PO21 2PN
Telephone 01243 842064

Index